EN-CRYPT
RISK ON, GAME
STRONG

GET READY TO NAVIGATE CYBERSECURITY RISKS IN THE ERA OF FINTECH PRODUCTS

HASHNEE SUBBUSUNDARAM &

BALASUNDARAM SUBBUSUNDARAM

Disclaimer

This book is for informational purposes only and does not constitute legal, financial, or professional advice. Readers should consult qualified professionals for guidance tailored to their specific needs. The authors and publishers are not liable for any errors, omissions, or outcomes resulting from the use of this content. References to regulations are informational and should not replace official sources. Use this material at your own discretion.

FOREWORD

"EN-CRYPT RISK ON, GAME STRONG is an intriguing and compelling read". I was intrigued not just by the blurb, but also the thorough, well-structured and well-compiled contents of the book.

The book contains a comprehensive coverage of Fin-Tech security, the uniqueness of risk factor associated and the concerns to be addressed while dealing with Fin-Tech Security".

PROF. V. KAMAKOTI
DIRECTOR, IIT MADRAS
Indian Institute of Technology Madras

What specifically caught my attention was the impact assessment owing to the confluence of technological advances like blockchain, quantum computing and AI ML on FinTech. The authors have comprehensively covered aspects of decentralized finance (a.k.a DeFi), which sooner or later could become the de facto framework of financial instruments and financial services. DeFi is often enabled through smart contracts on a programmable, permissionless blockchain. The authors address the regulatory concerns in a very interesting way, given the growing concern of implementing Anti-Money Laundering (AML) and Counter-Terrorism Financing (CTF) measures.

I have no hesitation in stating that this book by HASHNEE SUBBUSUNDARAM & BALASUNDARAM SUBBUSUNDARAM stands as a unique testament to their strength, lucidity in presenting a complex subject and their unwavering spirit. I am sure that this will resonate deeply with readers.

As Hashnee and Bala rightly pointed out, let us "GET READY TO NAVIGATE CYBERSECURITY RISKS IN THE ERA OF FINTECH PRODUCTS!".

TABLE OF CONTENTS

PART 1

THE INTERSECTION OF FINTECH AND CYBERSECURITY

CHAPTER 1
THE FINTECH REVOLUTION

📖 Introduction to the FinTech Revolution

The Financial Technology (FinTech) revolution has transformed the way we think about finance, making it more digital, efficient, and accessible than ever before. With each breakthrough, FinTech reshapes how individuals and businesses manage money, conduct transactions, and plan for the future. From seamless payment systems and automated wealth management to innovative insurance models and lending platforms, FinTech is revolutionizing every aspect of finance. Its promise? Faster, more transparent, and user-friendly financial services that meet the demands of a tech-savvy generation, redefining financial interactions across the globe.

Historically, the financial industry was slow to adopt new technology due to the complexities of legacy systems and the weight of strict regulatory frameworks. The traditional banking model, while reliable, often lacked the agility to respond to rapid changes in consumer behavior and technological advancement. Enter FinTech—a dynamic sector driven by agility, innovation, and customer-centric solutions. Today, FinTech companies are leading the charge, outpacing traditional banks and pushing the boundaries of digital innovation. Their rise signifies more than just technological advancement; it represents a shift in the financial paradigm itself, breaking down barriers to inclusion and empowering users to take control of their financial destinies.

At the heart of this revolution are cutting-edge technologies such as artificial intelligence (AI), machine learning, blockchain, and advanced data analytics. These tools have fundamentally redefined the scope of financial services, enabling products that are more scalable, adaptable, and intuitive. AI and machine learning power sophisticated algorithms capable of personalizing user experiences and optimizing decision-making processes. Blockchain technology promises unparalleled transparency and security in transactions, while advanced data analytics opens up new avenues for predictive insights and trend identification. Together, these technologies form the backbone of the FinTech industry, allowing it to challenge long-standing norms and set new standards in efficiency and innovation.

However, with great power comes great responsibility, and the exponential growth of FinTech brings its own set of challenges—chief among them, cybersecurity. As FinTech solutions rely heavily on digital platforms and interconnected systems, they have introduced a new era of cybersecurity risks. Every transaction, every data exchange, and every innovative feature represents a potential vulnerability.

For financial institutions and FinTech innovators alike, the stakes have never been higher. Cybersecurity is no longer just an operational concern—it is the cornerstone of trust, customer retention, and business continuity. As users increasingly entrust their financial data and transactions to digital platforms, ensuring the security of these interactions becomes paramount. Understanding the unique security risks posed by FinTech solutions is critical. From protecting sensitive user data and preventing financial fraud to navigating a labyrinth of global regulations, success in this competitive landscape requires vigilance, expertise, and a proactive approach.

The convergence of FinTech innovation and cybersecurity presents a complex yet crucial challenge. Strategies must be designed to safeguard innovation while building resilient and secure financial ecosystems. This involves not only the deployment of robust cybersecurity measures but also the fostering of a culture of security awareness across organizations. Collaboration among stakeholders—FinTech companies, regulators, and cybersecurity experts—is essential to establish standards that protect users without stifling innovation.

Ultimately, thriving in this environment demands a deep understanding of the intersection between technology and security. Only by addressing these challenges head-on can the FinTech industry continue to drive progress, fostering a future where financial services are not only innovative and efficient but also trusted and secure. As FinTech shapes the financial world of tomorrow, cybersecurity will remain the bedrock upon which its success is built.

Key Types of FinTech Products

1. Digital Wallets

Digital wallets, also known as e-wallets, have revolutionized how people manage money and conduct transactions. They enable users to store, send, and receive funds electronically, reducing the need for physical cash and cards. Popularized by companies like PayPal, Apple Pay, and Google Wallet, digital wallets have become indispensable in the era of contactless and mobile payments. These platforms allow users to link bank accounts, store card information, and make payments with just a few taps on their devices, transforming how we handle personal finance and everyday transactions.

Evolution and Growth of Digital Wallets

The evolution of digital wallets[1] can be traced back to the early days of e-commerce, where online payments were primarily processed through credit cards. PayPal was among the pioneers, simplifying online purchases for both buyers and merchants by providing a secure platform for financial transactions. As smartphones became more sophisticated, companies like Apple and Google integrated payment features into their devices, facilitating the adoption of mobile-based digital wallets and giving rise to a booming industry projected to reach $24.3 trillion by 2028 (Fortune Business Insights, 2023[2]).

Digital wallets are now used for various transactions, including P2P (peer-to-peer) transfers, online shopping, and even payments for in-store purchases via NFC (Near Field Communication) technology. The global shift toward a cashless society has only accelerated this trend, with many consumers preferring the convenience, speed, and accessibility of digital wallets over traditional payment methods.

Key Features of Digital Wallets

Digital wallets are designed to simplify and secure financial interactions. They allow users to:

- **Link Bank Accounts and Cards**: Users can link their bank accounts, credit cards, and debit cards to digital wallets, making it easy to transfer funds or make payments without manually entering financial information.

---

[1] https://www.finextra.com/blogposting/26756/the-rise-of-digital-wallets-trends-and-future-prospects
[2] https://www.fortunebusinessinsights.com/digital-payment-market-101972

- **Store Loyalty Cards and Coupons**: Many digital wallets enable users to store loyalty cards, rewards programs, and even digital coupons, enhancing the shopping experience by making these tools readily accessible at checkout.

- **Conduct Contactless Payments**: Digital wallets often use NFC technology to allow users to pay by simply holding their device near a contactless payment terminal, a feature that gained popularity during the COVID-19 pandemic as a hygienic payment method (Statista, 2021[3]).

- **Secure Authentication**: Many wallets include advanced authentication methods like fingerprint, face recognition, and PIN codes to secure access to the wallet, providing a layer of protection against unauthorized access.

Security Needs and Challenges

While digital wallets offer undeniable convenience, they also introduce security risks due to the sensitive financial and personal data they hold. Cybersecurity threats are constantly evolving, and digital wallets have become prime targets for cybercriminals aiming to exploit vulnerabilities. Major security challenges for digital wallets include:

1. **Data Breaches**: Digital wallets store sensitive data, including card information and bank account details. If breached, attackers could gain unauthorized access to a user's financial resources. Encrypting stored data is crucial to prevent data exposure, even in the event of a breach (Kaspersky, 2022).

2. **Weak Authentication Protocols**: Digital wallets rely on secure authentication methods to protect users' accounts.

---

[3] https://www.statista.com/topics/982/mobile-payments/

However, passwords and PINs can be vulnerable to brute-force attacks or social engineering schemes. Multi-factor authentication (MFA), which combines two or more verification factors (e.g., fingerprint, face recognition, or OTP), is widely regarded as a best practice for strengthening account security (Symantec, 2022).

3. **Phishing Attacks**: Phishing schemes are among the most common cyber threats targeting digital wallet users. Attackers use fake emails, messages, or websites to deceive users into sharing their login credentials or other sensitive information. To counter this, some digital wallets now deploy AI-based security tools to detect phishing links and block them before they reach users (Verizon, 2022).

4. **Tokenization and Encryption**: Digital wallets often employ tokenization, a process that replaces sensitive card details with unique tokens, reducing the risk of data exposure. Even if attackers intercept transaction data, the information remains useless without the actual card data. End-to-end encryption also ensures that all transmitted data is securely encoded, safeguarding it from unauthorized interception (McAfee, 2023).

5. **Continuous Security Updates**: With the sophistication of cyberattacks[4] increasing, digital wallets need constant updates to maintain security. Threat detection and vulnerability patches are regularly issued by developers to address new challenges and protect against the latest threats. Many digital wallets also work with cybersecurity

---

[4] https://lisnr.com/resources/blog/securing-digital-wallets-addressing-top-security-concerns-for-financial-services-leaders/

companies to conduct audits and test their security measures, enhancing user protection.

Emerging Security Innovations in Digital Wallets

To stay ahead of cybercriminals, digital wallet providers are investing in cutting-edge security technologies, including:

- **Behavioral Biometrics**: This emerging security measure analyzes users' typing patterns, finger pressure, and interaction habits with their devices to create a behavioral profile. Any deviation from the norm can trigger additional verification steps, adding a dynamic layer of security (IBM Security, 2023).

- **AI-Driven Threat Detection**: Artificial intelligence and machine learning are increasingly being used to detect potential security breaches. By analyzing patterns and anomalies in transaction data, AI can flag suspicious activities in real-time, helping to prevent unauthorized transactions before they occur (Gartner, 2023).

2. *Payment Gateways*

Payment gateways[5] have become essential in modern commerce, serving as intermediaries that securely facilitate online transactions between a buyer's bank and a merchant's bank. They allow businesses to accept digital payments globally, making them indispensable in the e-commerce sector. Pioneers like Stripe, Square, and PayPal have developed sophisticated payment solutions that seamlessly integrate into websites and mobile apps, enabling a frictionless payment experience for users. These platforms simplify transactions by supporting various payment methods, including

---

[5] https://thefinrate.com/the-importance-of-payment-gateways-for-e-commerce-success/

credit cards, bank transfers, and digital wallets, thereby enhancing convenience and accessibility for customers.

Beyond their front-end convenience, payment gateways provide the technical backbone that ensures accurate and efficient transaction processing. Key functionalities include transaction authorization, settlement, and detailed reporting, all of which are vital for managing high transaction volumes and providing a smooth user experience. In addition, payment gateways often include business management tools like currency conversion, invoicing, and tax calculation, which are particularly valuable for companies with a global customer base. As such, payment gateways are not only payment processors but also critical components of comprehensive e-commerce infrastructure.

Evolution and Growth of Payment Gateways

The concept of payment gateways originated in the early days of e-commerce when credit cards were the primary online payment method. PayPal was one of the first to streamline the online shopping experience by offering a secure platform that made it easier for both buyers and sellers to transact online. As technology advanced, new companies like Stripe and Square introduced sophisticated APIs and customization options, enabling businesses to embed payment processing directly into their websites and applications. This adaptability has fueled the rapid growth of the industry, with the global digital payment market projected to reach $24 trillion by 2028[6].

Today, payment gateways facilitate a wide array of transactions, including recurring billing, subscriptions, and in-app purchases, as well as global cross-border payments. The rise of mobile devices has further expanded the use of payment gateways, enabling users

---

[6] https://www.financemagnates.com/

to make quick, secure payments through their smartphones. The COVID-19 pandemic accelerated this trend as contactless payments became the preferred method of payment, increasing both the adoption and integration of payment gateways worldwide[7].

Security Needs and Challenges

With digital transactions on the rise, payment gateways face significant security challenges, as they handle high volumes of sensitive financial data. The primary security concerns include:

1. **Data Encryption:** Payment gateways must secure data transmission between the user's device and the bank's servers. Most gateways use Transport Layer Security (TLS) encryption to protect transaction data, preventing unauthorized interception.

2. **PCI-DSS Compliance:** The Payment Card Industry Data Security Standard (PCI-DSS) requires companies that handle card data to follow specific security protocols. This compliance is crucial to protect card data and prevent data breaches.

3. **Tokenization:** Sensitive payment details, such as card numbers, are often replaced with unique tokens, which are useless to attackers if intercepted. This minimizes data exposure and is an essential feature for protecting user data.

4. **Risk Scoring and Real-Time Monitoring:** Payment gateways use real-time analysis of transaction data to detect anomalies, flagging suspicious transactions before they are completed. This is often augmented by machine learning to adapt to evolving fraud patterns (Kshetri, 2020).

---

[7] https://pay10.com/blog-Evolution-of-Payment-Gateway-Integration.php

5. **Two-Factor Authentication (2FA):** Many payment gateways offer two-factor authentication, requiring users to verify their identity through a secondary method, such as a one-time passcode or biometric verification, adding an extra layer of protection.

Key Features of Payment Gateways

Modern payment gateways offer a suite of features designed to streamline and secure the transaction process. These include:

- **Transaction Authorization and Settlement:** Payment gateways verify transaction details and communicate with the issuing bank to ensure payment approval, enabling merchants to receive funds in their accounts.

- **Support for Multiple Payment Methods:** Most gateways accept payments via credit cards, debit cards, bank transfers, and digital wallets, catering to diverse user preferences.

- **Customization and Integration:** Payment gateways provide APIs and SDKs that enable businesses to customize the payment experience and seamlessly integrate it into their websites and apps.

- **Fraud Detection Tools:** Many payment gateways include built-in tools for fraud prevention, including monitoring for suspicious activity and employing risk scoring algorithms to flag potentially fraudulent transactions.

- **Detailed Reporting and Analytics:** Payment gateways offer reporting tools that allow businesses to analyze transaction data, helping them understand customer behavior and make data-driven decisions.

Emerging Security Innovations in Payment Gateways

To counter increasingly sophisticated cyber threats, payment gateway providers are investing in advanced security technologies, including:

- **Machine Learning for Fraud Detection:** AI and machine learning algorithms analyze large volumes of transaction data, detecting unusual patterns that may indicate fraud. This real-time detection helps to prevent fraudulent transactions and minimize losses (Gartner, 2023).

- **Biometric Verification:** Many payment gateways are implementing biometric authentication, such as fingerprint or facial recognition, which provides more secure and convenient access for users.

- **Blockchain Technology:** Some gateways are exploring blockchain for its transparent and immutable transaction records, which could enhance data security and reduce fraud in the long term (IBM Blockchain, 2023).

3. *Robo-Advisors*

Robo-advisors are digital platforms that use algorithms to provide automated financial planning and investment management services, often with minimal human involvement. They simplify the investment process for users, offering low-cost, customized portfolio management that aligns with individual risk preferences, financial goals, and timelines. Robo-advisors have democratized access to investment management by reducing fees and offering services to a broader range of investors, particularly those who may not have access to traditional financial advisors. Key players in the robo-advisory market include Betterment, Wealthfront, and

Vanguard's Digital Advisor, each known for providing tailored, automated investment solutions.

Robo-advisors typically follow a systematic, evidence-based approach to portfolio construction, leveraging principles of modern portfolio theory to diversify investments and optimize risk-adjusted returns. These platforms often rebalance portfolios automatically based on market conditions and user preferences, ensuring that users' portfolios remain aligned with their financial goals without requiring frequent manual adjustments. As a result, robo-advisors have become popular among novice and experienced investors alike, with global assets under management expected to exceed $2 trillion by 2025 as more individuals turn to these digital financial solutions (Statista, 2023[8]).

Evolution and Growth of Robo-Advisors

Robo-advisors emerged in the wake of the 2008 financial crisis as a low-cost alternative to traditional financial advisory services. They initially catered to millennial investors, who were more comfortable with digital solutions, lower fees, and self-directed financial planning. Over time, the technology has evolved, incorporating more sophisticated algorithms, machine learning, and artificial intelligence to provide increasingly personalized recommendations. Today's robo-advisors can offer a variety of investment options, including tax-loss harvesting, retirement planning, and socially responsible investing, making them suitable for a diverse range of investors.

The COVID-19 pandemic further accelerated the adoption of robo-advisors, as individuals sought digital financial services while traditional in-person advisory services faced limitations. As

---

[8] https://www.statista.com/forecasts/1259591/robo-advisors-managing-assets-united-states

consumer expectations shift towards digital-first solutions, robo-advisors have adapted by offering more intuitive user interfaces, educational resources, and even hybrid models that combine automated advice with human support. With continuous advancements in AI and financial technology, robo-advisors are expected to play an even larger role in personal finance, especially among tech-savvy investors and younger generations (Forbes, 2022[9]).

Security Needs and Challenges

Due to their handling of sensitive financial data and reliance on data-driven algorithms, robo-advisors face significant security challenges. Some of the primary security needs include:

1. **Data Privacy and Encryption:** Robo-advisors manage confidential data, including personal and financial information, which must be encrypted to prevent unauthorized access. Many robo-advisors use robust encryption protocols to secure both stored and transmitted data.

2. **Algorithm Integrity and Transparency:** Given that robo-advisors use algorithms to make investment decisions, it's essential that these algorithms are secure and transparent. Manipulated algorithms could lead to improper investment advice, undermining user trust. Ensuring algorithm integrity and explaining investment logic to users is crucial to maintain transparency and trust.

3. **Secure Authentication Mechanisms:** Multi-Factor Authentication (MFA) is widely used to protect user accounts, requiring multiple verification steps (e.g., one-

---

[9] http://www.forbes.com/sites/robertberger/2015/02/05/7-robo-advisors-that-make-investing-effortless/

time passcodes, biometrics) to prevent unauthorized access. This additional layer of security reduces the risk of account compromise.

4. **Protection Against Cyber Attacks:** With cyber threats constantly evolving, robo-advisors are targets for attacks aimed at accessing sensitive financial data. Firewalls, intrusion detection systems, and regular security audits are necessary to identify and mitigate vulnerabilities.

5. **Compliance with Data Privacy Regulations:** Robo-advisors must comply with data protection laws, such as the General Data Protection Regulation (GDPR) and the California Consumer Privacy Act (CCPA), which mandate strict controls over data collection, storage, and sharing. Adherence to these regulations ensures that user data is managed responsibly and mitigates the risk of legal repercussions (Kaspersky, 2023[10]).

Key Features of Robo-Advisors

Robo-advisors provide various features designed to enhance the user experience and streamline investment management. These include:

- **Automated Portfolio Management:** Robo-advisors automatically build and adjust portfolios based on user preferences, using algorithms to optimize asset allocation for better returns.

- **Low-Cost Investment Options:** By reducing overhead costs and minimizing human involvement, robo-advisors offer lower fees compared to traditional advisory services, making investment more accessible.

---

[10] https://www.kaspersky.com/blog/gdpr-questions-answered/23097/

- **Risk Assessment and Goal Setting:** Robo-advisors assess each user's risk tolerance and financial goals, ensuring that their investment strategy aligns with their risk preferences.

- **Tax Optimization:** Many robo-advisors offer tax-efficient investing options, such as tax-loss harvesting, to help users minimize their tax burden and increase after-tax returns.

- **User-Friendly Interface:** Robo-advisors are designed for ease of use, often featuring simple interfaces that guide users through account setup, portfolio tracking, and financial planning.

Emerging Security Innovations in Robo-Advisors

To safeguard sensitive data and enhance user trust, robo-advisors are adopting new security technologies, including:

- **Behavioral Analytics for Fraud Detection:** Robo-advisors are increasingly using behavioral analytics to detect suspicious activity by analyzing users' typical patterns and flagging deviations, such as unusual login locations or unusual withdrawal requests.

- **AI-Driven Threat Detection:** Artificial intelligence helps identify anomalies in transaction patterns and detect potential security breaches in real time. AI-based threat detection can prevent unauthorized transactions and alert users to potential fraud.

- **Blockchain Technology for Data Security:** Some robo-advisors are exploring blockchain technology for its potential to create immutable, secure transaction records that offer transparency and reduce the risk of data tampering (IBM, 2023).

4. *Blockchain Platforms*

Blockchain technology has emerged as a revolutionary force in the financial technology (FinTech) sector, particularly within cryptocurrency and decentralized finance (DeFi) applications. At its core, blockchain offers decentralized and transparent systems that eliminate the need for traditional intermediaries like banks, reducing transaction costs and improving efficiency. By using distributed ledger technology (DLT), blockchain enables secure, peer-to-peer transactions that are cryptographically verified and immutable. This technology has formed the backbone of numerous applications, from cryptocurrency platforms like Bitcoin and Ethereum to smart contract-based DeFi projects.

Popular blockchain platforms such as Bitcoin, Ethereum, and Ripple have gained widespread adoption for their ability to provide secure and verifiable transactions. Bitcoin, the first cryptocurrency, introduced the concept of a decentralized, permissionless ledger. Ethereum expanded on this concept by integrating smart contracts—self-executing contracts with the terms directly written into code—enabling decentralized applications (dApps) and further expanding the blockchain's use cases. Ripple, another major player, focuses on enabling fast, low-cost cross-border payments and has gained recognition for its innovative consensus algorithm, which reduces the energy consumption associated with traditional proof-of-work mechanisms.

The benefits of blockchain are vast, ranging from enhanced transparency and traceability to greater security and fraud prevention. However, as blockchain adoption grows, it introduces new security challenges that must be addressed to ensure the integrity and safety of transactions.

Evolution and Growth of Blockchain Platforms

Blockchain platforms have seen rapid evolution and growth since the inception of Bitcoin in 2008. The initial focus on cryptocurrencies as a store of value and medium of exchange has expanded to include a wide array of use cases. Ethereum's introduction in 2015 paved the way for decentralized applications (dApps) and smart contracts, which have become foundational elements of the decentralized finance (DeFi) ecosystem. Today, Ethereum is moving toward a more scalable, energy-efficient model through the transition to Ethereum 2.0, which incorporates proof-of-stake (PoS) in place of its traditional proof-of-work (PoW) mechanism.

The growth of blockchain platforms has also led to the development of layer 2 solutions, which aim to enhance scalability and transaction throughput without compromising on security. Platforms like Polygon, Optimism, and Arbitrum have emerged as solutions that build on top of Ethereum's existing infrastructure, improving transaction speeds and reducing gas fees, while maintaining decentralization and security.

Furthermore, the rise of central bank digital currencies (CBDCs) and the increased interest in institutional adoption of blockchain technology for secure payments and asset management have contributed to the widespread recognition of blockchain's potential. With projections of blockchain adoption in global financial services growing exponentially, its market size is expected to surpass $160 billion by 2029 (Grand View Research, 2023[11]).

---

[11] https://www.grandviewresearch.com/industry-analysis/blockchain-technology-market

Security Needs and Challenges

Despite its inherent advantages, blockchain technology is not immune to cybersecurity threats. As blockchain platforms continue to grow in popularity, several key security concerns have emerged:

1. **Securing Wallets and Private Keys:** Blockchain wallets store the private keys needed to access users' cryptocurrency holdings. If these keys are compromised, attackers can gain control over the funds. Ensuring that private keys are securely stored (e.g., using hardware wallets) is crucial to maintaining the integrity of blockchain transactions.

2. **Smart Contract Vulnerabilities:** While smart contracts automate transactions and agreements, they are also vulnerable to bugs and exploits. If a smart contract is poorly written or contains flaws in its code, it can be exploited by malicious actors. One notable example of this was the DAO hack in 2016, where a vulnerability in the code was exploited to drain millions of dollars from the Ethereum network.

3. **51% Attacks:** A 51% attack occurs when a malicious actor gains control of more than 50% of the network's mining or staking power, enabling them to rewrite the blockchain's transaction history and potentially double-spend funds. Although proof-of-work (PoW) systems like Bitcoin are highly secure, smaller blockchains with lower hash rates are more vulnerable to such attacks.

4. **Sybil Attacks:** In Sybil attacks, attackers create multiple fake identities to gain disproportionate control over the network. These types of attacks are particularly concerning in decentralized networks where consensus mechanisms rely on the participation of nodes.

5. **Transaction Verification Integrity:** Ensuring the validity and authenticity of transactions is critical. Blockchain platforms use consensus algorithms (e.g., proof-of-work, proof-of-stake) to achieve agreement among distributed nodes, but any vulnerability in these mechanisms can lead to fraudulent transactions or data manipulation.

6. **Regulatory Compliance and Privacy:** Blockchain platforms, particularly those related to cryptocurrencies, face challenges in complying with regulatory frameworks like Anti-Money Laundering (AML) and Know Your Customer (KYC) regulations. Additionally, ensuring user privacy while maintaining transparency is a delicate balance that needs to be addressed by many blockchain platforms.

Key Features of Blockchain Platforms

Blockchain platforms offer various features that set them apart from traditional financial systems, including:

- **Decentralization:** Unlike traditional centralized systems, blockchain operates on a decentralized network of nodes, removing the need for intermediaries and enabling peer-to-peer transactions. This decentralization increases transparency and reduces the risk of single points of failure.

- **Security and Cryptography:** Blockchain platforms use advanced cryptographic techniques, including hashing and public-private key pairs, to secure transactions and protect user identities. This makes it nearly impossible to alter transaction records without being detected.

- **Immutability:** Once a transaction is recorded on a blockchain, it is immutable and cannot be altered or deleted.

This feature is critical for ensuring the integrity and traceability of data within the network.

- **Smart Contracts:** Smart contracts enable the automation of agreements and transactions based on predefined conditions. These self-executing contracts eliminate the need for intermediaries and reduce the risk of fraud or human error.

- **Transparency and Audibility:** Every participant on a blockchain network has access to the same data, ensuring that transactions are transparent and verifiable. This enhances trust and accountability within the network.

Emerging Security Innovations in Blockchain Platforms

To address these security challenges, blockchain platforms are adopting innovative security measures:

- **Layer 2 Solutions:** These solutions, including state channels and rollups, help increase blockchain scalability and transaction speeds while maintaining a high level of security. By offloading some transactions from the main chain, these solutions reduce congestion and improve the efficiency of blockchain networks.

- **Zero-Knowledge Proofs (ZKPs):** Zero-knowledge proofs allow one party to prove to another that they know a value without revealing the value itself. This technique enhances privacy by enabling secure transactions without exposing sensitive data, and it is being increasingly integrated into blockchain platforms for both security and privacy.

- **Post-Quantum Cryptography:** As quantum computing advances, the cryptographic algorithms used by current blockchain systems may become vulnerable to decryption. To prepare for this, blockchain developers are exploring post-quantum cryptography methods to secure blockchain transactions against future quantum attacks.

- **Decentralized Identity Management:** To address privacy and regulatory concerns, decentralized identity (DID) solutions are being implemented on blockchain networks. DIDs allow individuals to control their personal data without relying on centralized authorities, reducing the risk of data breaches.

5. *Decentralized Finance (DeFi)*

Decentralized Finance (DeFi) is an innovative subset of blockchain-based finance that aims to disrupt traditional financial services by eliminating intermediaries such as banks, brokers, and payment processors. DeFi platforms leverage blockchain technology and smart contracts to create decentralized financial ecosystems where users can engage in peer-to-peer transactions, lending, borrowing, trading, and more, all without relying on centralized authorities. By offering open and permissionless access, DeFi enables anyone with an internet connection to access financial services, regardless of their location or socioeconomic status.

DeFi projects, including Compound, Uniswap, and Aave, are leading the charge in offering user-controlled and transparent financial solutions. Compound allows users to lend and borrow cryptocurrencies, while Uniswap is a decentralized exchange (DEX) that facilitates trustless token swaps. Aave, another prominent DeFi platform, offers decentralized lending services, enabling users to earn interest or take out loans using their crypto assets as collateral. By using blockchain technology, DeFi platforms ensure that all transactions are recorded on an immutable ledger, creating a high level of transparency and security that is not typically found in traditional finance.

DeFi represents a paradigm shift, emphasizing openness, accessibility, and decentralization. However, this nascent industry is not without its challenges, particularly in terms of security. As DeFi platforms grow in popularity and scale, they face significant cybersecurity threats that need to be addressed to protect users' funds and maintain trust in the ecosystem.

Evolution and Growth of DeFi

The DeFi movement[12] began gaining traction in 2018, with the launch of projects like MakerDAO, which introduced the concept of decentralized lending through the use of the DAI stablecoin. However, the sector truly took off in 2020, when the total value locked (TVL) in DeFi protocols surged from just $1 billion in January 2020 to over $80 billion by the end of the year. This growth was fueled by the increasing demand for decentralized alternatives to traditional financial products, such as lending, insurance, and asset management.

The rise of Automated Market Makers (AMMs) like Uniswap played a pivotal role in DeFi's growth by allowing users to trade tokens directly from their wallets, without the need for intermediaries. Meanwhile, platforms like Compound and Aave introduced decentralized lending and borrowing markets, enabling users to earn interest on their crypto holdings or take out loans by using crypto as collateral. As the DeFi ecosystem expanded, new protocols were developed to offer decentralized derivatives, insurance products, and even prediction markets, further diversifying the financial services available.

---

[12] https://www.statista.com/statistics/1272181/defi-tvl-in-multiple-blockchains/

DeFi's open-source nature has led to rapid innovation and development, but also to increased risk. The lack of centralized control means that users must take responsibility for their own security, including managing private keys and ensuring the integrity of smart contracts. As of 2023, the DeFi market has seen continued growth, with hundreds of billions of dollars in TVL and a rapidly expanding user base. However, with this growth comes increased attention from regulators and the need for a secure, scalable infrastructure.

Security Needs and Challenges

While DeFi offers numerous advantages, its decentralized and open-source nature creates several security challenges that need to be addressed:

1. **Smart Contract Vulnerabilities:** The most significant risk in DeFi arises from vulnerabilities in smart contracts. If a smart contract contains coding errors or bugs, malicious actors can exploit these weaknesses to steal funds or manipulate the contract's behavior. High-profile hacks, such as the DAO attack in 2016, have demonstrated the risks associated with smart contracts. To mitigate these risks, DeFi platforms often conduct code audits and engage third-party security firms to identify and patch vulnerabilities before they are exploited.

2. **Liquidity Pool Risks:** DeFi platforms often rely on liquidity pools, where users provide funds in exchange for rewards or fees. However, liquidity pools can be exposed to "impermanent loss," where the value of the assets within the pool changes relative to the market price. Additionally, flash loan attacks—where attackers borrow large amounts of capital in a single transaction to exploit vulnerabilities—

pose a significant threat to DeFi platforms. Proper risk management strategies and liquidity monitoring are essential to mitigate these risks.

3. **Decentralized Governance Risks:** While decentralized governance is a key feature of DeFi, it can also present security risks. If a platform's governance token becomes concentrated in the hands of a few entities, they could potentially take control of the platform and manipulate decisions in their favor. Additionally, governance attacks, such as vote manipulation or Sybil attacks, could undermine the integrity of the platform's decision-making process.

4. **User Security and Privacy:** DeFi platforms do not have customer support or centralized entities to assist users who lose access to their wallets or fall victim to phishing attacks. Users must take full responsibility for the security of their private keys, which are crucial for accessing funds. As such, ensuring secure wallet storage and providing robust educational resources on phishing prevention is essential for user safety.

5. **Regulatory Risks:** DeFi platforms often operate in regulatory gray areas, and governments are beginning to examine the potential for enforcing traditional financial regulations on decentralized platforms. Issues like anti-money laundering (AML) and know-your-customer (KYC) compliance are particularly challenging in DeFi, where pseudonymous transactions are the norm. The lack of regulatory oversight can also make it harder for users to recover stolen funds or hold bad actors accountable.

Key Features of DeFi Platforms

DeFi platforms provide various features that set them apart from traditional financial services:

- **Peer-to-Peer Transactions:** DeFi platforms allow individuals to interact directly with one another, eliminating the need for intermediaries like banks or financial institutions. This facilitates fast, low-cost, and borderless transactions.

- **Smart Contracts:** DeFi relies heavily on smart contracts—self-executing contracts with the terms of the agreement directly written into code. These contracts automatically enforce the terms of the agreement, such as transferring funds or collateral, based on pre-determined conditions.

- **Decentralized Governance:** Many DeFi platforms are governed by Decentralized Autonomous Organizations (DAOs), which enable users to participate in decision-making processes regarding the platform's rules and updates. Token holders can vote on proposals, ensuring a more democratic system of governance.

- **Lending and Borrowing:** DeFi platforms like Compound and Aave allow users to lend their crypto assets to earn interest or borrow funds by using their crypto holdings as collateral. This opens up new ways of generating passive income and accessing liquidity.

- **Decentralized Exchanges (DEXs):** Platforms like Uniswap provide decentralized trading, allowing users to swap tokens without relying on a centralized exchange. These platforms use Automated Market Makers (AMMs) to set prices and provide liquidity.

- **Stablecoins:** Many DeFi platforms use stablecoins, such as DAI and USDC, which are pegged to a stable asset like the US dollar. Stablecoins are crucial for providing a stable store of value in the volatile cryptocurrency market, especially for lending and borrowing applications.

Emerging Security Innovations in DeFi

To address these security challenges, DeFi platforms are investing in various innovations:

- **Code Audits and Formal Verification:** Many DeFi platforms now undergo rigorous smart contract code audits to identify potential vulnerabilities before deployment. Some platforms are also adopting formal verification techniques, which mathematically prove the correctness of a smart contract's code, further reducing the risk of vulnerabilities.

- **Decentralized Insurance:** DeFi platforms are exploring decentralized insurance models to protect users against risks such as smart contract bugs, hacks, or losses in liquidity pools. These insurance protocols allow users to purchase coverage directly through smart contracts, providing an additional layer of protection.

- **Oracles for Secure Data Feeds:** Oracles are trusted external data providers that supply smart contracts with real-world information, such as asset prices or market data. By integrating secure and reliable oracles, DeFi platforms can ensure that smart contracts execute based on accurate and up-to-date information, reducing the risk of manipulation.

- **Privacy Enhancements:** To protect user privacy while maintaining transparency, DeFi platforms are adopting privacy-enhancing technologies such as zero-knowledge proofs (ZKPs), which allow users to prove the validity of a transaction without revealing sensitive information.

📖 Conclusion

The FinTech revolution has fundamentally transformed how we manage, invest, and spend money. From digital wallets to DeFi platforms, each type of FinTech product has its own unique features and specific security needs. As cyber threats become more sophisticated, the security measures protecting these innovations need to keep up. By tackling these challenges with proactive and well-rounded strategies, the FinTech industry can keep growing and thriving in our ever-evolving digital world.

💡 What We Learnt:

- **FinTech's Big Impact**: FinTech has completely reshaped finance by bringing in digital solutions that make banking and investing more accessible, efficient, and centered around you—the consumer.

- **A Variety of FinTech Tools**: FinTech isn't just one thing; it's a range of products like digital wallets, payment gateways, robo-advisors, blockchain, and decentralized finance (DeFi). Each of these tools has its own special features and security needs.

- **Unique Security Needs**: With each FinTech product comes specific cybersecurity risks, from data breaches and weak passwords to phishing scams and smart contract vulnerabilities.

- **Digital Wallets**: These digital wallets are convenient for storing financial info, but keeping them safe requires encryption, multi-factor authentication, and tokenization to secure sensitive data.

- **Payment Gateways**: When it comes to payment gateways, security is a top priority. Encryption, PCI-DSS compliance, tokenization, and fraud detection help ensure that transactions are safe and secure.

- **Robo-Advisors**: These automated investment platforms need solid data privacy, secure algorithms, and strong user authentication to make sure your investments are both safe and reliable.

- **Blockchain and DeFi**: Blockchain and DeFi bring decentralized, peer-to-peer transactions, but they face unique challenges like smart contract bugs, 51% attacks, and compliance with regulations.

CHAPTER 2
WHY CYBERSECURITY IS CRUCIAL IN FINTECH

The world of FinTech, where finance meets cutting-edge technology, is an exciting place. It's fast, innovative, and always looking for the next big thing. But as thrilling as it sounds, there's one huge concern that can't be overlooked—cybersecurity. In an industry that revolves around money and sensitive personal data, cybersecurity isn't just important; it's absolutely essential. Let's dive into why cybersecurity is so crucial in FinTech and what makes it such a unique challenge in this space.

📖 Building User Trust and Protecting Data

At the core of every FinTech business lies a fundamental element: user trust. Consider this—would you readily share your bank account details, personal information, or transaction history with a company you didn't completely trust? Most individuals would hesitate. In the FinTech sector, trust is as invaluable as any tangible asset. It represents an implicit agreement that says, "Your data and finances will be handled with the utmost care." This trust is critical because users need confidence that their money and personal information are secure, safeguarded by robust cybersecurity measures.

Imagine an individual discovering a promising app for budgeting or investing. While they may hear glowing recommendations from friends, any whispers of past data breaches or the absence of visible security protocols can cause hesitation. A prime example is the 2017 Equifax data breach[13], which exposed sensitive information of over 147 million individuals. Although Equifax wasn't a FinTech company, the incident reverberated through both the finance and technology sectors, highlighting the vulnerabilities in data protection. It served as a stark reminder for companies across industries, including FinTech, to prioritize security to safeguard their reputations and user confidence.

For FinTech companies, establishing trust often begins with transparency. A startup launching a mobile payment solution, for instance, might include a comprehensive explanation within their app detailing how user data is managed and protected. Companies like PayPal and Venmo have set industry benchmarks by implementing robust encryption methods and offering real-time alerts for suspicious activities. When organizations communicate openly about their security practices and demonstrate a commitment to safeguarding user data, they foster a sense of reliability and encourage users to engage with their services.

In addition to building trust, FinTech companies must prioritize the equally critical responsibility of data protection. These companies handle vast amounts of sensitive personal and financial information, ranging from credit card numbers and account balances to detailed transaction histories. This data represents a lucrative target for cybercriminals, making the FinTech industry particularly vulnerable. A single data breach can have devastating consequences for users, exposing them to fraud, identity theft, and other significant risks. For FinTech companies, safeguarding this data is not just a sound business practice—it is a legal imperative. Regulations such

---

[13] https://www.strongdm.com/what-is-equifax-data-breach

as the General Data Protection Regulation (GDPR) in Europe and the California Consumer Privacy Act (CCPA) enforce stringent standards for data management, with noncompliance resulting in substantial penalties. However, beyond regulatory adherence, robust data protection demonstrates a company's genuine commitment to its users.

Robinhood serves as a notable example of a FinTech company addressing data protection with urgency. Following an incident in 2020 where over 2,000 user accounts were compromised due to unauthorized access, the company promptly implemented enhanced security measures, including two-factor authentication. Additionally, Robinhood actively educated its users on the importance of adopting strong security practices. While the breach was unfortunate, the company's swift response and visible improvements played a crucial role in maintaining user trust[14].

In the digital age, data security is not merely a task on a checklist—it is a fundamental element of a successful FinTech company's operations. Some startups incentivize users to adopt stronger security measures, such as biometric logins, by offering rewards like discounts or cash-back programs. Others collaborate with leading cybersecurity firms to strengthen their defenses, assuring users that their data is protected by experts. By prioritizing data protection, FinTech companies not only mitigate risks but also distinguish themselves in a competitive market, showcasing their commitment to safeguarding user information.

Ultimately, trust and data protection are inextricably linked for FinTech companies striving for long-term success. For users, knowing that a company prioritizes security allows them to focus on the benefits of the service without concern for potential

---

[14] https://www.bbc.com/news/technology-59209494

vulnerabilities. When considering a new FinTech app or service, evaluating the security measures in place is essential. Companies that prioritize trust and data protection demonstrate they are deserving of user confidence and loyalty.

📖 The Unique Cybersecurity Challenges in FinTech

In today's digital era, cybersecurity is a critical concern across industries, but in the FinTech sector, the stakes are exceptionally high. FinTech companies are transforming how we manage money, make payments, and access financial services through their digital-first approach. While these innovations bring immense convenience, they also introduce unique cybersecurity challenges that demand constant vigilance and adaptation. Let's explore what makes cybersecurity in FinTech distinct and the risks these companies must navigate to safeguard their platforms and users.

1. *High Volume and Speed of Transactions*

One hallmark of FinTech is the sheer volume and speed of its transactions. Unlike traditional financial institutions that often process transactions in batches or with delays, FinTech operates in real-time. Hundreds of transactions occur every second, akin to a bustling highway where efficiency and precision are paramount. However, this also creates countless potential entry points for cybercriminals.

🔍 **Example:** Peer-to-peer payment apps like Venmo or Cash App facilitate instant money transfers, making financial interactions seamless. Yet, every transaction must be authenticated while simultaneously flagging suspicious activity. Imagine a hacker compromising one account and initiating a rapid series of transactions to funnel out funds. Without robust real-time detection systems, such vulnerabilities can result in significant financial losses. Protecting every dollar in motion while maintaining a smooth user experience requires advanced security infrastructure capable of acting in milliseconds.

2. Handling Sensitive Personal Data

For FinTech companies, data is the lifeblood of their operations—valuable both to their business and to cybercriminals. Bank account numbers, credit card details, social security numbers, and other personally identifiable information (PII) are prime targets for hackers who can exploit them for identity theft, fraud, or sale on the dark web.

🔍 **Example:** Consider digital wallets like Apple Pay or Google Pay. Users entrust these platforms with their most sensitive financial details, expecting the same level of security as a physical bank vault. A single breach could expose millions of users' data, causing reputational damage and financial harm. To counter this, companies invest heavily in encryption, tokenization, and multi-factor authentication. These measures ensure that even if hackers breach a system, the data they access is indecipherable and unusable.

3. Rapid Innovation: A Double-Edged Sword

FinTech thrives on constant innovation, racing to deliver user-friendly features and seamless experiences. However, this pace of development often comes with security trade-offs. Each new feature, app update, or software tweak can unintentionally introduce vulnerabilities.

🔍 **Example:** The cryptocurrency boom has spurred the development of platforms like Binance and Coinbase, offering users tools to trade digital assets. Yet, with innovation comes risk. In 2022, several cryptocurrency exchanges faced devastating hacks, resulting in losses exceeding millions of dollars. These incidents underscore the importance of rigorous security testing and threat modeling during every phase of development. Balancing cutting-edge innovation with robust security measures isn't just a best practice—it's a necessity.

4. *Navigating Complex Regulatory Landscapes*

The FinTech sector operates within an intricate web of regulations that vary across regions and evolve constantly. From the General Data Protection Regulation (GDPR) in Europe to the California Consumer Privacy Act (CCPA) in the United States, companies must adhere to stringent standards for collecting, storing, and processing user data.

🔍 **Example:** Imagine a FinTech platform serving users in Europe, North America, and Asia. To remain compliant, it must implement data protection measures tailored to each region, such as encryption standards in line with GDPR and opt-out mechanisms for CCPA compliance. Failure to comply not only risks hefty fines—like the €50 million penalty Google faced under GDPR—but also erodes user trust. For smaller firms, these challenges are magnified as they balance limited resources with the need for global compliance[15].

5. *Insider Threats and Third-Party Risks*

The collaborative nature of FinTech often involves partnerships with third-party vendors for services like payment processing, cloud storage, and customer support. While these relationships drive scalability, they also introduce vulnerabilities. Each vendor represents a potential weak link, especially if their security measures are insufficient.

🔍 **Example:** In 2020, a large FinTech firm faced a breach originating from a third-party vendor, exposing thousands of customer records. Such incidents emphasize the need for stringent vendor assessments and ongoing security audits. Internally, insider threats also pose risks. An employee's accidental click on a phishing email can compromise entire systems. Comprehensive training programs and role-based access controls are vital to mitigating these threats[16].

Looking Ahead

As FinTech continues to innovate, the cybersecurity landscape will grow even more complex. Companies must prioritize proactive measures—like integrating AI for threat detection, conducting regular security audits, and fostering a culture of security awareness. By striking a balance between innovation and protection, FinTech

---

[15] https://www.bitdefender.com/en-us/blog/hotforsecurity/google-fined-50-million-eur-for-violating-gdpr-rules#:~:text=Google%20Fined%2050%20Million%20EUR%20for%20Violating%20GDPR%20Rules

[16] https://www.forbes.com/sites/larsdaniel/2024/11/20/global-fintech-giant-finastra-investigating-data-breach/

firms can not only safeguard their platforms but also build the trust essential for long-term success.

Cybersecurity in FinTech isn't just about preventing breaches; it's about preserving the integrity of the financial ecosystem in a digital-first world.

📖 Navigating Cybersecurity Regulations

In the world of FinTech, securing user data is more than just a box to check—it's a cornerstone of trust between companies and their customers. As cyber threats evolve, so does the regulatory landscape, pushing companies to adopt stringent measures to safeguard personal and financial information. If you're in FinTech or just curious about the industry, understanding these regulations can help shed light on how companies protect your data. Let's dive into some of the major regulations shaping cybersecurity and why complying with them is essential for any forward-thinking FinTech firm.

PCI DSS (Payment Card Industry Data Security Standard)

The Payment Card Industry Data Security Standard (PCI DSS)[17] is a set of security protocols established to protect cardholder information during online transactions. Whenever you make an online purchase, PCI DSS ensures that your card details are safeguarded against potential threats. All entities involved in processing, storing, or transmitting card data are required to adhere

---

[17] https://en.wikipedia.org/wiki/Payment_Card_Industry_Data_Security_Standard

to these standards, which aim to secure every card transaction and mitigate the risk of data breaches.

For FinTech companies, compliance with PCI DSS is not optional but mandatory. The standard outlines requirements such as encrypting cardholder data, maintaining secure networks, and continuously monitoring systems for vulnerabilities. Beyond reducing fraud risks, adherence to PCI DSS fosters customer confidence by demonstrating a commitment to securing payment information.

🔍 Real-World Examples:

1. **Target Data Breach (2013):** In 2013, Target experienced a significant data breach where hackers accessed the credit and debit card information of approximately 40 million customers. The breach was attributed to vulnerabilities in Target's payment system, highlighting the critical importance of PCI DSS compliance in preventing such incidents[18].

2. **Heartland Payment Systems Breach (2008):** Heartland Payment Systems, a payment processing company, suffered a data breach in 2008 that compromised over 100 million card numbers. The breach was due to malware that intercepted card data during the transaction process, underscoring the necessity for robust security measures as prescribed by PCI DSS[19].

---

[18] https://coverlink.com/cyber-liability-insurance/target-data-breach/
[19] https://www.proofpoint.com/us/blog/insider-threat-management/throwback-thursday-lessons-learned-2008-heartland-breach

These incidents underscore the critical role of PCI DSS in safeguarding cardholder data and the severe consequences of non-compliance.

GDPR (General Data Protection Regulation)

The General Data Protection Regulation (GDPR)[20], introduced by the European Union in 2018, has redefined global standards for data privacy, influencing how businesses manage user information. One of its most noticeable impacts is the ubiquitous cookie consent banners on websites, ensuring that users are informed about data collection practices. However, GDPR extends far beyond cookies— it mandates businesses to communicate transparently about how they collect, store, and process personal data.

For **FinTech companies**, GDPR compliance represents both a regulatory requirement and a strategic opportunity. For instance, under GDPR, users have the right to access their data, know how it is being used, and request its deletion. Companies like **Revolut** and **PayPal**, which serve European users, must implement robust measures to protect user data, even if their headquarters are outside the EU. This involves steps such as encrypting sensitive information, obtaining explicit consent for data processing, and ensuring users can easily manage their privacy preferences. Real-world cases illustrate the significance of GDPR compliance.

🔍 For example, **Google was fined €50 million in 2019** for failing to provide clear and accessible information about its data processing practices, highlighting the financial and reputational risks of non-compliance (CNIL,

---

[20](Sources: CNIL, 2019; European Commission GDPR Documentation)

2019[21]). Conversely, companies that align with GDPR can foster trust, as demonstrated by startups like **N26**, which prominently feature their data protection policies to reassure users.

In a broader context, GDPR compliance is a commitment to user privacy at a time when data breaches are rampant. By adhering to these standards, FinTech companies signal transparency and accountability, distinguishing themselves in a competitive market. Beyond legal obligations, this approach enhances customer loyalty, as users increasingly prefer platforms that prioritize their privacy and security.

By embracing GDPR as more than a compliance challenge, FinTech firms can strengthen their brand reputation while safeguarding users in an era of heightened data awareness.

PSD2 (Payment Services Directive 2)

The Revised Payment Services Directive (PSD2)[22] has significantly transformed electronic payments within the European Union by enhancing security measures and fostering innovation. A cornerstone of PSD2 is the implementation of Strong Customer Authentication (SCA), which mandates multi-factor authentication (MFA) for electronic transactions. This requirement ensures that electronic payments are performed with multi-factor authentication, increasing the security of electronic payments.

Key Aspects of PSD2 and MFA:

---

[21] https://gdprhub.eu/index.php?title=CNIL_(France)_-_SAN-2019-001#:~:text=The%20CNIL%20imposed%20a%20record,and%20the%20transparent%20information%20principle.

[22] https://financialit.net/news/security-and-compliance/psd2-and-mfa-what-exactly-do-they-mean

- **Strong Customer Authentication (SCA):** SCA requires the use of at least two out of three authentication elements:

- **Knowledge:** Something only the user knows (e.g., password or PIN).

- **Possession:** Something only the user possesses (e.g., mobile phone or hardware token).

- **Inherence:** Something the user is (e.g., fingerprint or facial recognition).

For instance, when accessing an online banking account, a user might be required to enter a password (knowledge) and confirm a code sent to their mobile device (possession).

1. **Impact on FinTech Companies:** FinTech firms must integrate SCA into their services to comply with PSD2. This involves updating authentication processes to include MFA, thereby enhancing security and maintaining competitiveness. For example, a mobile payment app may implement fingerprint recognition alongside traditional passwords to meet SCA requirements.

2. **Exemptions and Flexibility:** While SCA is mandatory for most transactions, PSD2 allows certain exemptions to ensure user convenience without compromising security. Low-value transactions, recurring payments, and transactions deemed low-risk by the payment service provider may be exempt from SCA. This flexibility helps balance security needs with user experience.

3. **Open Banking and Innovation:** PSD2 promotes open banking by requiring banks to provide third-party providers (TPPs) access to customer account information, with the customer's consent. This has led to the development of new

financial services and applications, such as budgeting tools and payment initiation services, enhancing customer choice and fostering innovation in the financial sector[23].

🔍 Examples of PSD2 Implementation:

- **Banking Applications:** Many European banks have updated their mobile applications to include biometric authentication methods, such as fingerprint or facial recognition, in addition to traditional passwords, to comply with SCA requirements.

- **E-commerce Platforms:** Online retailers have integrated MFA into their payment processes. Customers may be prompted to enter a one-time password sent to their mobile device when making a purchase, ensuring that the transaction is authorized by the legitimate account holder.

By enforcing these measures, PSD2[24] aims to create a more secure and user-friendly environment for electronic payments, reducing fraud and building trust among consumers and businesses alike.

📖 Why Compliance Is a Competitive Edge

Following these regulations isn't just about avoiding fines or keeping up with legal standards; it's a way to win the trust of increasingly informed and cautious customers. People today are more aware of data privacy risks and want to feel confident that their information is being handled responsibly. A FinTech company that

---

[23] https://www.riskified.com/learning/ecommerce-enablement/psd2-explained-a-guide-to-the-not-so-new-payment-services-directive/

[24] https://www.gpayments.com/about/psd2-strong-customer-authentication-3d-secure-2/

proudly demonstrates PCI DSS, GDPR, and PSD2 compliance is signaling to its users that it takes cybersecurity seriously, which, in turn, can set it apart from competitors.

In an industry where trust is paramount, regulatory compliance acts as a badge of reliability. It assures customers that a company has gone the extra mile to protect their data, giving them peace of mind with every transaction. By staying compliant, FinTech companies aren't just meeting requirements; they're building a stronger, more transparent relationship with their customers, who, in turn, are more likely to stay loyal.

In the fast-paced FinTech world, cybersecurity regulations are only set to evolve, reflecting the growing complexities of cyber threats. Staying compliant and transparent isn't just good practice—it's essential for any FinTech company aiming to thrive in today's market.

📖 Why FinTech Cybersecurity Differs from Traditional Finance

Cybersecurity within the FinTech space isn't a simple replica of what we see in traditional finance. Although they both handle sensitive financial data and are targets for cyber threats, the approach to security in FinTech is distinct due to the nature and demands of this rapidly growing sector. Let's dig into the key reasons why FinTech's cybersecurity needs differ, what makes it more complex, and some real-life examples that highlight these differences.

1. The "Digital-First" Foundation

Most FinTech companies are built with a "digital-first" approach, meaning their products and services are often accessible online, on mobile devices, and frequently incorporate emerging technologies. Unlike traditional banks, which have slowly transitioned from physical to digital over decades, FinTech companies typically start out fully digital, which means cybersecurity is central from day one.

Traditional banks, such as Chase or Bank of America, may use online and mobile platforms, but these are layered onto existing security systems that were originally designed for physical bank branches. FinTech, however, has no such physical constraints; it was built for speed and agility online. As a result, FinTech companies need to be adaptable and proactive, updating their cybersecurity strategies as quickly as technology—and threats—evolve.

> 🔍 **Example**: Think about PayPal or Square. These platforms offer instant, mobile transactions and are used widely on smartphones, making them highly convenient but also attractive targets for cybercriminals. To counter these threats, they have to employ sophisticated encryption and fraud detection systems to maintain user trust in a way that's scalable and efficient for their digital infrastructure.

2. Limited Resources and Fast Growth

While established banks have decades of resources and infrastructure, many FinTech companies are startups. They may be launching with limited funds and smaller teams, making it harder to invest in top-notch cybersecurity from the start. Traditional financial institutions have the time and resources to build comprehensive security frameworks and regularly update them, while FinTech companies must find quick, cost-effective solutions that won't slow down their growth.

🔍 **Example**: Robinhood, the popular investment app, faced security challenges as it grew rapidly, offering trading with minimal fees. In 2020, the company faced a breach where hackers targeted some user accounts. This highlighted the balancing act for FinTech companies between scaling fast and ensuring security. Robinhood's response included ramping up security, improving customer service, and educating users, all under the pressure of rapid growth[25].

3. Higher Expectations and Scrutiny

FinTech is disrupting the finance world, which brings a lot of attention from regulators, investors, and users. They're under constant scrutiny to uphold security standards that are at least equal to, if not more rigorous than, those of traditional banks. Even though many FinTech companies are still young, they're held to standards typically expected of established financial institutions.

Unlike traditional banks, which have longstanding relationships with regulators and customers, FinTechs often have to prove themselves from scratch. Customers and regulators are wary of the risk involved, as many have heard stories of cyber threats like phishing or identity theft that could compromise their financial information.

🔍 **Example**: In Europe, the Revised Payment Services Directive (PSD2) regulation requires strong customer authentication for electronic payments, which places

---

[25] https://therecord.media/robinhood-discloses-security-breach-and-extortion-attempt

pressure on FinTech companies to implement multi-factor authentication and other robust security measures. Monzo, a digital bank based in the UK, has had to adopt these regulations to assure its customers that their accounts are secure, aligning its security efforts with expectations typically reserved for long-established banks.

4. Innovation-Driven and Highly Integrated Systems

FinTech companies often partner with or build upon other digital platforms and services to deliver integrated financial experiences. While this interconnectedness can make services seamless for users, it also opens up more vulnerabilities, as security gaps in any single system can potentially compromise others.

For instance, many FinTech apps and platforms utilize APIs (Application Programming Interfaces) to interact with other services—like linking a bank account to a budgeting app. While APIs are essential for creating a cohesive user experience, they're also vulnerable points if not properly secured, as they could provide hackers with entry points into multiple interconnected systems.

🔍 **Example**: Venmo, a popular payment service, enables users to transfer funds easily by connecting to their bank accounts. The simplicity of integration with various platforms makes it attractive to users, but also to hackers who may try to exploit these connections. In response, Venmo and similar companies implement strong authentication, real-time transaction monitoring, and other cybersecurity measures to keep these integrations safe.

5. User Education and Empowerment

While traditional banks are working to educate users on cybersecurity best practices, FinTech companies often put even more emphasis on this aspect, given their fully digital environments and younger user bases. Users of digital financial services need to understand the risks of using online platforms and take personal steps to protect their accounts, like enabling two-factor authentication and being cautious of phishing schemes.

🔍 **Example**: Cash App, used by millions for peer-to-peer payments, constantly prompts users to verify any changes in their accounts and reminds them about security best practices. The platform educates users on identifying potential fraud attempts and encourages setting up PIN codes for extra security, underscoring the shared responsibility of securing personal information.

📖 Learning from Real-World Breaches

The FinTech industry, known for its rapid innovation and user-friendly financial solutions, has unfortunately faced numerous high-profile cybersecurity breaches over the years. Each of these incidents serves as a stark reminder of the necessity for strong, proactive security measures. Let's dive into a few noteworthy examples and discuss what both companies and users can learn from them.

1. *Revolut: Safeguarding User Trust in a Digital World*

One of the more alarming breaches in recent years involved Revolut[26], a popular FinTech app with a massive user base across the globe. In 2022, the platform experienced a data breach that compromised the sensitive information of over 50,000 customers. Thankfully, no financial details were reportedly stolen, but personal data, including names and contact details, was exposed. While Revolut quickly responded by tightening security measures and improving user notification protocols, the incident was a wake-up call. For users, it highlighted the reality that even trusted, widely-used FinTech apps are not immune to breaches, emphasizing the importance of enabling features like two-factor authentication to add an extra layer of security.

Imagine relying on an app to handle your finances, only to discover that your private information has been exposed. It's a situation no user wants to find themselves in, yet it's a possibility that looms over every digital platform. This incident reminds both companies and users to be vigilant—no app, no matter how established, is ever truly "breach-proof."

2. *Robinhood: When User Information Becomes a Target*

Another headline-grabbing breach occurred in 2021 with Robinhood[27], a popular trading platform that allows users to trade stocks and cryptocurrency with ease. In this case, a hacker gained access to sensitive information, including millions of users' email addresses and full names. Though no financial losses were reported,

---

[26] https://techcrunch.com/2022/09/20/revolut-cyberattack-thousands-exposed/
[27] https://newsroom.aboutrobinhood.com/robinhood-announces-data-security-incident-update/

the breach highlighted a different kind of threat: the exposure of personally identifiable information (PII), which could be exploited for phishing scams or social engineering attacks.

For Robinhood, this was a moment to strengthen security protocols and work on regaining user trust. For users, it was a reminder to be cautious with unsolicited messages or emails, especially those appearing to come from familiar companies. As a best practice, FinTech users should regularly update passwords, remain skeptical of unexpected communications, and familiarize themselves with the signs of phishing attempts.

3. Cash App: Insider Threats and the Risks from Within

Cybersecurity isn't just about keeping outsiders out—it's also about monitoring those on the inside. In 2022, Cash App[28] experienced a breach not from an external hacker, but from a former employee who downloaded sensitive information related to customers' brokerage accounts. This was a prime example of an "insider threat," which can be especially difficult to predict and prevent.

This incident underscored the importance of internal security protocols, especially for employees who have access to sensitive data. FinTech companies are now placing greater emphasis on monitoring employee activities and limiting access based on role-specific needs. As for users, this breach highlighted the value of regularly reviewing account activity, enabling alerts for any suspicious actions, and reaching out to customer support if anything unusual arises. The breach was also a reminder to think twice before

---

[28] https://www.techradar.com/news/cash-app-alerts-8-million-customers-to-data-breach

sharing sensitive information—even with trusted employees or representatives.

🔍 Common Threads and Lessons Learned

Each of these breaches, while unfortunate, has provided valuable lessons for both companies and users alike. FinTech companies are recognizing the need to adopt a multi-layered approach to security, focusing not just on external defenses but also on internal protocols and user education. Many are now launching user education campaigns, aimed at promoting basic cybersecurity practices, like enabling two-factor authentication, creating strong passwords, and avoiding common phishing traps.

For users, these breaches reinforce the importance of being vigilant. The convenience of FinTech apps shouldn't overshadow the need for personal security measures. Users should take the time to enable all available security features, stay informed about the latest cybersecurity threats, and treat all unexpected communications with caution.

📖 Moving Forward: Building a Safer FinTech Ecosystem

In response to these incidents, FinTech companies are investing in advanced security measures. Many are adopting artificial intelligence (AI) and machine learning (ML) tools that can detect unusual activity patterns, helping to prevent breaches before they occur. Some companies are also offering cybersecurity workshops or tips through their apps, empowering users to take a more active role in their security.

🔍 For example, companies like Square have implemented mandatory security alerts and step-by-step guides for users to set up stronger authentication. Meanwhile, smaller FinTech startups are exploring innovative ways to secure data, such as encrypted biometrics and behavioral analysis, that can more accurately identify legitimate users and catch potential intruders.

📖 Conclusion

Cybersecurity in FinTech is both a challenge and an opportunity. While there's no denying the hurdles—regulations, evolving threats, and the demand for rapid innovation—there's also the chance to create an industry that users can trust. By prioritizing cybersecurity, FinTech companies can build a foundation of trust, protect their users, and set themselves apart in an increasingly crowded market.

In the end, as FinTech continues to grow, so too must its focus on cybersecurity. Only then can it deliver on its promise of revolutionizing finance safely and responsibly.

📖 The Cybersecurity Imperative for FinTech

As FinTech continues to reshape the global financial landscape, cybersecurity has emerged as a foundational element crucial to sustaining trust, growth, and operational integrity. The digital transformation in the financial services industry, driven by innovations such as digital wallets, payment gateways, robo-advisors, blockchain platforms, and decentralized finance (DeFi) products, has introduced new and complex cybersecurity risks. The interconnected nature of these technologies means that vulnerabilities in one system can lead to cascading security breaches across the entire financial ecosystem.

Cybersecurity Risks in the FinTech Sector

FinTech firms handle vast amounts of sensitive financial data, making them prime targets for cybercriminals. These companies are often faced with Advanced Persistent Threats (APTs), ransomware attacks, and phishing schemes. A single breach can not only lead to financial losses but can also severely damage an organization's reputation and erode consumer trust. With innovations like blockchain and DeFi disrupting traditional financial services, the scope of potential attacks has expanded.

    🔍   For example, attacks on smart contracts or vulnerabilities in Decentralized Exchanges (DEXs) can have far-reaching consequences, as evidenced by several high-profile hacking incidents in the DeFi space over the past few years. Moreover, with the growing dependence on cloud infrastructure and third-party service providers, the risk of supply chain attacks has also increased, as cybercriminals exploit weaknesses in these external partners' security postures[29].

To mitigate these risks, FinTech companies must adopt dynamic and proactive cybersecurity measures. This includes leveraging state-of-the-art tools such as artificial intelligence-based threat detection systems that can identify and neutralize emerging threats in real-time. These systems are becoming increasingly important as cybercriminals use sophisticated techniques, including machine learning, to bypass traditional security measures. Additionally, Multi-Factor Authentication (MFA) has become an industry-

---

[29] https://www.weforum.org/stories/2020/07/great-reset-fintech-financial-technology-cybersecurity-controls-cyber-resilience-businesses-consumers/

standard safeguard against unauthorized access, ensuring that even if a user's password is compromised, additional verification steps prevent further exploitation.

Regulatory and Compliance Challenges

Alongside technological solutions, regulatory compliance remains one of the most significant challenges for FinTech companies. As the industry grows and expands into new markets, the need for consistent adherence to global and regional standards has become paramount. Regulatory bodies are under increased pressure to ensure that FinTech companies uphold stringent data protection standards, maintain financial transparency, and prevent illicit activities such as money laundering and fraud.

The General Data Protection Regulation (GDPR) in the European Union and the Payment Card Industry Data Security Standard (PCI-DSS) are among the key regulatory frameworks that set strict guidelines for how FinTech companies must handle consumer data. The Financial Action Task Force (FATF) guidelines also mandate compliance with anti-money laundering (AML) regulations, further emphasizing the need for robust cybersecurity measures to detect and prevent illegal transactions. Failure to comply with these regulations can result in hefty fines and legal repercussions, alongside damage to a company's credibility[30].

Furthermore, new regulations are emerging globally, making it essential for FinTech firms to stay informed and agile in adapting to the evolving regulatory landscape. For instance, recent discussions around the digital euro and the regulatory frameworks for Central Bank Digital Currencies (CBDCs) are expected to have significant implications on FinTech companies dealing with digital currencies and blockchain technology. Therefore, regulatory bodies are

---

[30] https://eur-lex.europa.eu/legal-content/EN/TXT/?uri=LEGISSUM%3A310401_2

working hand-in-hand with FinTech companies to ensure both innovation and security go hand in hand.

Balancing Innovation and Security

The challenge for FinTech companies lies in finding the right balance between innovation and security. On one hand, the sector is driven by a desire to innovate—whether it's enhancing user experience, creating new financial products, or tapping into emerging markets. On the other hand, the very nature of these innovations can create unforeseen vulnerabilities. For instance, with the rise of open banking, where third-party providers are granted access to consumer financial data, comes the increased risk of exposing sensitive information to malicious actors if not properly secured[31].

As FinTech expands into new markets, maintaining this delicate balance becomes even more challenging. A rapid rollout of new services or technologies may compromise security, which in turn undermines consumer confidence. The key to overcoming this challenge is embedding security into the core of FinTech product design and development processes. This approach, often referred to as "security by design," integrates security considerations from the inception of a project, ensuring that security features are built into the product's architecture rather than added as an afterthought.

User Education and Internal Policies

In addition to technological solutions and regulatory compliance, user education is a critical component of cybersecurity strategy in the FinTech space. Many cyberattacks, such as phishing scams or social engineering attacks, target end-users rather than systems or software. Therefore, educating consumers on the importance of

---

[31] https://www.openbanking.exchange/wp-content/uploads/economist-open-banking-report.pdf

strong password hygiene, recognizing phishing attempts, and understanding the risks of unsecured networks is essential for reducing the likelihood of breaches.

Equally important is the development of robust internal cybersecurity policies. FinTech companies must establish comprehensive guidelines for their employees, from conducting regular security training to implementing strict access controls and conducting periodic security audits. Insider threats, whether intentional or inadvertent, remain a significant concern for FinTech firms, as employees or contractors with access to sensitive data can pose considerable risks. By fostering a culture of security awareness and vigilance, FinTech companies can better protect themselves from both external and internal threats.

Partnerships with Cybersecurity Firms

As FinTech companies continue to grow and diversify, many are realizing the importance of forming strategic partnerships with established cybersecurity firms. These partnerships provide FinTech companies with access to specialized expertise, threat intelligence, and advanced security tools that may otherwise be difficult or expensive to develop in-house. Cybersecurity firms can offer services such as vulnerability assessments, penetration testing, and continuous monitoring to ensure that FinTech systems remain resilient against evolving threats.

Furthermore, collaborative efforts between financial institutions, regulatory bodies, and cybersecurity experts can foster a more secure digital ecosystem. By sharing threat intelligence, best practices, and insights, these stakeholders can work together to create stronger defenses against emerging cyber threats.

💡 What We Learnt:

- **User Trust is Essential**: FinTech companies need to build strong user trust through transparent practices and robust cybersecurity to ensure users feel safe sharing personal and financial data.

- **High Frequency of Transactions**: The fast-paced nature of real-time transactions in FinTech creates numerous entry points for potential cyber threats.

- **Handling Sensitive Data**: FinTech companies handle highly sensitive data, including bank and personal information, making them prime targets for cybercriminals.

- **Rapid Innovation and Vulnerability**: The constant introduction of new features and updates in FinTech opens up new vulnerabilities that hackers could exploit.

- **Complex Regulatory Compliance**: FinTech companies must adhere to regulations like PCI DSS, GDPR, and PSD2, which add layers of security but also operational complexity.

- **Insider and Third-Party Risks**: Collaborations with third parties and employee access to sensitive data create additional security challenges for FinTech firms.

- **PCI DSS Compliance**: Essential for companies handling card transactions, PCI DSS provides guidelines to secure transactions and enhance user trust.

- **GDPR and User Privacy**: GDPR mandates transparency in data handling, demonstrating FinTech companies' commitment to respecting user privacy.

- **PSD2 and Multi-Factor Authentication**: PSD2's requirement for multi-factor authentication improves transaction security and builds user confidence.

- **Proactive User Education**: FinTech companies educate users on security best practices, promoting a shared responsibility approach to cybersecurity.

- **Why Cybersecurity Matters**: As FinTech grows, so does the need for cybersecurity. Constant updates, real-time threat detection, and sticking to regulatory standards are essential for managing risks.

- **Balancing Innovation and Security**: FinTech companies have to balance rapid innovation with strong security practices to maintain user trust and keep their platforms safe.

- **User Education and Policies**: Educating users on best cybersecurity practices and creating strong company policies can help reduce risks from social engineering and insider threats.

- **Working with Cybersecurity Experts**: Partnering with cybersecurity specialists gives FinTech companies access to the latest tools, threat insights, and best practices to stay one step ahead of cyber threats.

- **The Future of FinTech Security**: FinTech security will need to keep evolving, adapting proactively to stay resilient against increasingly complex cyber threats.

PART 2
COMMON CYBERSECURITY CHALLENGES IN FINTECH

CHAPTER 3
OVERVIEW OF CYBERSECURITY THREATS

The rapid adoption of FinTech has transformed financial services, but it has also expanded the cybersecurity threat landscape. As innovation progresses, so do the strategies of cybercriminals seeking to exploit vulnerabilities. This chapter explores the various types of cybersecurity threats in FinTech, attack vectors, and specific risks associated with key FinTech categories such as digital wallets, payment gateways, and robo-advisors.

📖 Types of Cybersecurity Threats in FinTech

The FinTech ecosystem, an amalgamation of finance and technology, continues to grow exponentially, but with growth comes heightened exposure to cyber risks. Cybersecurity threats in this domain can have far-reaching consequences, including financial loss, reputational damage, legal ramifications, and diminished trust among customers. Here's an in-depth look at these threats:

Fraud and Identity Theft

Fraud and identity theft are among the most persistent cybersecurity challenges faced by the FinTech industry. Cybercriminals[32] exploit

---

[32]https://fintechmagazine.com/fraud-id-verification/phishing-one-of-the-most-common-security-threats-in-banking

vulnerabilities in digital platforms to steal sensitive financial and personal information, such as social security numbers, credit card details, and login credentials. These details are often obtained through phishing emails, fake websites, or malware.

Once criminals gain access to sensitive data, they impersonate legitimate users to conduct unauthorized transactions, open accounts, or manipulate financial systems. Fraud detection systems can mitigate this risk, but attackers constantly innovate ways to bypass these mechanisms.

A significant contributor to fraud is credential stuffing, where attackers use previously stolen login details from other breaches to access accounts. The financial toll of fraud and identity theft has been rising steadily, with an estimated $41 billion lost globally in 2022 alone[33]. Awareness and the adoption of multi-factor authentication are essential in combating this threat.

Data Breaches

Data breaches[34] represent a critical threat to the FinTech sector, exposing confidential user and company data to unauthorized parties. Such breaches occur due to weak encryption, misconfigured systems, or targeted attacks like SQL injections. In addition to financial harm, breaches severely impact a company's reputation, leading to loss of customer trust and regulatory penalties.

FinTech startups are especially vulnerable due to their reliance on cloud technologies and third-party APIs, which, while enhancing scalability and functionality, also introduce new attack vectors. The sensitive nature of FinTech data—ranging from customer account

---

[33] https://www.juniperresearch.com/press/ecommerce-losses-online-payment-fraud-48bn/
[34] https://arxiv.org/abs/2312.01752

details to payment records—makes breaches particularly devastating.

   🔍 For example, in 2020, the FinTech app Dave suffered a massive breach that exposed personal data of over 7 million users, highlighting the urgent need for robust security frameworks. Companies must adopt encryption protocols, implement access controls, and regularly audit their systems to minimize risks[35].

Ransomware Attacks

Ransomware attacks[36] have surged as one of the most destructive forms of cyberattacks. In these incidents, attackers encrypt critical company data and demand a ransom—often in cryptocurrency—in exchange for decryption keys. The encrypted data may include customer account information, transaction records, or even business continuity plans, effectively paralyzing operations.

In 2021, the banking sector reported a significant increase in ransomware incidents, with attackers exploiting remote working arrangements and poorly secured systems[37]. Beyond financial losses, these attacks tarnish a company's reputation and can lead to regulatory scrutiny.

The adoption of backup solutions, real-time monitoring, and endpoint protection tools is critical for mitigating the impact of ransomware. Organizations must also invest in employee training to recognize phishing attempts, a common entry point for ransomware.

---

[35] https://www.pymnts.com/news/security-and-risk/2020/fintech-dave-data-breach-hackers/
[36] https://arxiv.org/abs/2311.14783
[37] https://www.cisa.gov/news-events/cybersecurity-advisories/aa22-040a

Distributed Denial-of-Service (DDoS) Attacks

DDoS attacks are orchestrated attempts to overwhelm an organization's servers or networks with massive amounts of traffic, rendering systems inoperable. These attacks are often used as a smokescreen for other malicious activities, such as data theft or fraud, and can lead to significant financial and operational disruptions[38].

In the FinTech sector, where real-time transactions and 24/7 availability are critical, even a brief outage caused by a DDoS attack[39] can erode customer trust. For instance, in 2022, a well-known digital payment platform experienced a DDoS attack that left users unable to complete transactions for hours.

To defend against DDoS attacks, companies must invest in scalable cloud-based solutions, deploy intrusion prevention systems, and implement traffic filtering to handle abnormal surges in traffic.

Insider Threats

Insider threats[40] stem from individuals within the organization—employees, contractors, or business partners—who have access to sensitive information and systems. These threats can be intentional (e.g., disgruntled employees stealing data) or unintentional (e.g., employees inadvertently clicking on phishing links).

Detecting insider threats is particularly challenging due to the legitimate access held by the individual. For instance, an employee misusing privileged access to siphon funds or sharing customer data

---

[38] https://www.upguard.com/blog/biggest-cyber-threats-for-financial-services

[39] https://gcore.com/learning/ddos-attack-on-fintech/

[40] https://yellow.systems/blog/cybersecurity-in-fintech

with external entities can severely harm an organization's credibility.

Companies must foster a culture of security awareness, conduct regular audits, and employ user behavior analytics (UBA) tools to detect anomalies in access patterns. Additionally, Role-Based Access Control (RBAC)[41] ensures that employees only have access to the data they need for their roles.

📖 Critical Threats in FinTech Security: Phishing, Malware, and Social Engineering

As the FinTech industry continues to revolutionize the way financial services are accessed and delivered, it also faces a significant challenge: cyber threats. Among the most common and damaging attack vectors are phishing, malware, and social engineering. Each of these exploits not only targets technological vulnerabilities but often capitalizes on human error, making them particularly difficult to guard against. Below is an in-depth exploration of these threats, their implications for FinTech, and how organizations can mitigate them.

Phishing

Phishing attacks are among the most prevalent cyber threats in FinTech. These attacks typically involve fraudulent emails, text messages, or social media communications that appear to come from trusted entities. The goal is to trick users into divulging sensitive information such as login credentials, credit card numbers, or bank account details.

---

[41] https://www.upguard.com/blog

🔍 For example, an attacker might impersonate a financial institution and send an email asking users to "verify their account" via a malicious link. The success of phishing attacks often lies in their ability to exploit trust and urgency, leading individuals to act without fully scrutinizing the source of the communication. According to the *Anti-Phishing Working Group*, phishing attacks surged globally in 2022, with a record-breaking 1.2 million reported incidents in the first quarter alone. (APWG, 2022)[42]

Malware

Malicious software, or malware, poses another significant threat to FinTech systems. This type of software is designed to infiltrate digital devices and systems, enabling attackers to steal sensitive information, monitor transactions, or even disrupt entire operations. Common types of malware used in FinTech attacks include keyloggers, ransomware, and Trojans. For instance, ransomware can lock financial institutions out of their systems until a ransom is paid, often in cryptocurrency. Malware attacks frequently exploit outdated software, unpatched systems, or insufficient endpoint security, making regular updates and strong cybersecurity protocols critical. A *2023 report by Cybersecurity Ventures* estimates that damages from ransomware alone could reach $30 billion globally by the end of the year[43]. (Cybersecurity Ventures, 2023)

---

[42] https://www.stationx.net/phishing-statistics/
[43] https://cybersecurityventures.com/most-ransomware-attacks-occur-when-security-staff-are-asleep/

Social Engineering

Social engineering attacks leverage psychological manipulation to deceive individuals into revealing confidential information or granting access to secure systems. Unlike phishing, which relies on deceptive messages, social engineering often involves direct interaction, such as a phone call or in-person request.

> 🔍 For example, an attacker impersonating an IT support technician to gain access to sensitive company systems. These attacks can bypass even the most advanced technical safeguards by exploiting the human element, which is often the weakest link in security. Research from *IBM's Cyber Security Intelligence Index* reveals that human error is a contributing factor in 95% of cybersecurity breaches, highlighting the critical role of employee training and awareness[44]. (IBM, 2023)

FinTech's Unique Vulnerabilities

FinTech companies are particularly vulnerable to these attack vectors due to their heavy reliance on digital platforms and innovative technologies. While the industry's rapid pace of innovation offers convenience and accessibility to consumers, it can sometimes outpace the implementation of robust security measures. This creates opportunities for cybercriminals to exploit weaknesses, whether through unpatched software vulnerabilities, inadequate user authentication protocols, or insufficient employee training on cybersecurity best practices[45].

---

[44] https://newsroom.ibm.com/2023-07-24-IBM-Report-Half-of-Breached-Organizations-Unwilling-to-Increase-Security-Spend-Despite-Soaring-Breach-Costs

[45] https://bawn.com/risk-resilience-bawns-guide-to-cybersecurity-and-beyond/7-cybersecurity-challenges-facing-fintechs-and-small-financial-institutions-in-2025

Mitigating the Risks

To combat these threats, FinTech organizations must adopt a multi-layered approach to cybersecurity. This includes implementing advanced threat detection systems, conducting regular employee training to recognize and respond to phishing and social engineering attempts, and maintaining a robust incident response plan. Additionally, organizations should invest in the continuous updating of their systems and software to stay ahead of emerging threats. Collaboration within the industry and with regulatory bodies can also play a crucial role in developing standardized security practices and sharing threat intelligence.

By understanding the nuances of phishing, malware, and social engineering, FinTech companies can better protect themselves and their customers from potentially devastating breaches. Security must remain a top priority as the industry continues to grow and innovate, ensuring trust and resilience in the digital financial ecosystem.

📖 Common Vulnerabilities in FinTech

FinTech platforms are revolutionizing the financial industry, offering unparalleled convenience, speed, and innovation. However, these advancements come with unique security challenges. The interconnected nature of FinTech systems, combined with their reliance on digital technologies, opens up multiple avenues for cyber threats. To ensure trust and resilience, it is critical to address the following vulnerabilities:

API Exploits

Application Programming Interfaces (APIs)[46] are the backbone of FinTech platforms, enabling the seamless integration of services and functionalities. However, unsecured APIs can serve as entry points for attackers, exposing sensitive data and system controls.

🔍 For example, a report by Salt Security highlighted a critical vulnerability in a U.S.-based FinTech platform's API, which could have allowed attackers to take over administrative accounts. This underscores the need for continuous API monitoring, the use of rate limiting, and robust encryption to safeguard data.

Weak Authentication Mechanisms

Authentication serves as the first line of defense in protecting FinTech platforms from unauthorized access. Weak authentication[47] methods, such as simple or reused passwords, increase vulnerability to brute-force attacks and credential theft. The Open Worldwide Application Security Project (OWASP) recommends the implementation of multi-factor authentication (MFA) and adaptive authentication systems to mitigate these risks. A failure to strengthen authentication mechanisms can result in large-scale breaches, eroding user trust and financial losses.

---

[46] https://salt.security/press-releases/salt-security-discovers-critical-api-security-vulnerability-that-would-have-enabled-administrative-account-takeover-on-fintech-platform-serving-hundreds-of-banks
[47] https://www.akamai.com/site/en/documents/white-paper/2024/api-security-in-financial-services-mitigating-risks-and-ensuring-trust.pdf

Unpatched Software

Outdated software and systems lacking critical security patches are among the most exploited vulnerabilities in FinTech. Cybercriminals actively target these systems to deploy malware, ransomware, or steal sensitive financial data. Regular updates and a robust patch management policy are essential to close these security gaps. The Financial Conduct Authority (FCA) emphasizes that organizations must maintain updated systems to ensure operational resilience and compliance with regulatory standards.

Third-Party Risks

The interconnected nature of FinTech means that platforms often rely on third-party vendors to provide additional services, such as payment processing or cloud storage. While these collaborations can enhance functionality, they also introduce risks if third-party vendors do not follow stringent cybersecurity protocols. A comprehensive vendor risk management strategy, including regular audits and contractual obligations, is crucial to mitigate these threats. According to Ncontracts, third-party risks[48] include operational, reputational, and compliance-related vulnerabilities.

📖 Specific Threats in Key FinTech Categories

In the dynamic realm of financial technology (FinTech), various sectors encounter unique security challenges that necessitate vigilant attention. Below is an in-depth exploration of specific threats within key FinTech categories:

---

[48] https://www.ncontracts.com/nsight-blog/top-10-risks-third-party-vendors-pose-financial-institutions

Digital Wallets:

1. **Data Breaches:** Digital wallets store sensitive information such as card details and transaction histories, making them attractive targets for cybercriminals. Unauthorized access can lead to significant financial losses and identity theft. For instance, researchers have uncovered security issues in popular digital wallets like Apple Pay and Google Pay, highlighting vulnerabilities that could be exploited by attackers[49].

2. **Phishing and Social Engineering:** Attackers often employ deceptive tactics, such as fake alerts or websites that mimic legitimate wallet services, to trick users into revealing personal information. These schemes can compromise account security and lead to unauthorized transactions. A notable example is the rise in phishing scams targeting digital wallet users, where criminals send fraudulent messages to obtain sensitive data[50].

3. **Tokenization Flaws:** While tokenization enhances security by replacing sensitive data with unique identifiers, improper implementation can introduce vulnerabilities. Flaws in the tokenization process may expose critical information to unauthorized parties. It's essential to ensure that tokenization methods are robust and comply with industry standards to prevent such issues[51].

Payment Gateways:

1. **Man-in-the-Middle Attacks:** These attacks involve intercepting unencrypted data between users and payment

---

[49] https://www.fintechnews.org/researchers-uncover-security-issues-in-digital-wallets/
[50] https://www.news.com.au/finance/commonwealth-banks-move-to-protect-digital-wallets-as-scams-cost-aussies-208m/news-story/cd6b319cec573434ab4a4cee50b01596?
[51] https://en.wikipedia.org/wiki/Tokenization_%28data_security%29

processors, potentially compromising transaction integrity and exposing sensitive information. Ensuring robust encryption protocols is essential to mitigate this risk. For example, vulnerabilities in payment gateways can be exploited by attackers to intercept and manipulate transaction data[52].

2. **PCI DSS Non-Compliance:** The Payment Card Industry Data Security Standard (PCI DSS) sets requirements for securing cardholder data. Non-compliance can lead to data breaches, financial penalties, and reputational damage. Adherence to these standards is crucial for maintaining trust and security. Organizations that fail to comply with PCI DSS may face significant consequences, including legal actions and loss of customer confidence[53].

3. **Fraudulent Transactions:** Cybercriminals continually develop sophisticated schemes to exploit vulnerabilities in payment gateways, facilitating unauthorized fund transfers and financial fraud. Implementing advanced fraud detection mechanisms is vital to counteract these threats. The increasing complexity of fraud schemes necessitates continuous monitoring and updating of security measures to protect against potential breaches[54].

Robo-Advisors:
1. **Algorithm Manipulation:** Robo-advisors rely on algorithms to provide financial recommendations. Malicious actors may attempt to exploit weaknesses in these algorithms, leading to incorrect advice and potential

---

[52] https://en.wikipedia.org/wiki/Man-in-the-browser
[53] https://www.mdpi.com/2078-2489/11/12/590
[54] https://www.pwc.in/assets/pdfs/beyond-the-cloud-navigating-fintech-cyber-threats-and-fortifying-defences.pdf

financial losses for users. Ensuring the integrity and security of these algorithms is essential to maintain user trust and the effectiveness of robo-advisory services[55].

2. **Unauthorized Access:** Weak account security measures can result in hackers gaining access to user portfolios, allowing them to alter investment strategies or withdraw funds without authorization. Strengthening authentication protocols is essential to protect user assets. Implementing multi-factor authentication and regular security audits can help mitigate this risk[56].

3. **Data Privacy Breaches:** Handling sensitive financial data necessitates robust encryption and strict adherence to data protection laws. Breaches can lead to the exposure of personal information, undermining user trust and potentially resulting in legal consequences. For instance, inadequate data protection measures in robo-advisory services can result in significant privacy violations and financial losses for users[57].

📖 Addressing the Threat Landscape

As the FinTech sector continues to grow, so does its appeal to cybercriminals seeking to exploit vulnerabilities in digital financial systems. To safeguard sensitive data, ensure regulatory compliance, and maintain customer trust, FinTech companies must take a proactive stance against emerging cyber threats. The following measures are critical in building a robust security framework:

---

[55] https://www.emerald.com/insight/content/doi/10.1108/IJBM-10-2022-0439/full/html
[56] https://en.wikipedia.org/wiki/Multi-factor_authentication
[57] https://gsconlinepress.com/journals/gscarr/sites/default/files/GSCARR-2024-0241.pdf

Multi-Factor Authentication (MFA)

Implementing Multi-Factor Authentication[58] is one of the most effective ways to strengthen user account security. By requiring multiple forms of verification—such as a combination of passwords, biometrics, or one-time codes—MFA significantly reduces the risk of unauthorized access. This layered approach to authentication ensures that even if one factor is compromised, the system remains secure. Moreover, advancements in biometric authentication, such as fingerprint and facial recognition, provide seamless yet highly secure user experiences.

Regular Audits and Penetration Testing

Frequent security audits and penetration testing are essential for identifying potential vulnerabilities in systems and applications before they can be exploited. These assessments help organizations uncover weak points in their defenses, simulate real-world cyberattacks, and implement corrective measures. By embedding this practice into their operational workflows, FinTech companies can stay ahead of threats and demonstrate a commitment to cybersecurity excellence[59].

AI-Driven Threat Detection

Artificial intelligence (AI) has emerged as a game-changer in the fight against cybercrime. By analyzing vast amounts of transaction data in real time, AI-powered tools can identify suspicious activities and potential fraud with remarkable accuracy.

---

[58] https://www.velmie.com/post/2fa-for-fintech-best-practices
[59] https://www.bpm.com/insights/fintech-risk-management/

🔍 For example, AI can flag unusual login patterns, detect fraudulent transactions, or even predict potential threats based on historical behavior. Integrating AI-driven threat detection not only enhances security but also minimizes disruptions to legitimate users by reducing false positives.

User Education and Awareness

While sophisticated technologies are crucial, the human element remains a critical line of defense. Empowering users through education ensures they can recognize and respond to common cyber threats, such as phishing attacks or social engineering tactics. Regular training sessions, engaging tutorials, and timely reminders about security best practices can significantly reduce the likelihood of user-related breaches. For instance, users should be encouraged to create strong passwords, enable security features like MFA, and remain vigilant against unsolicited communications[60].

Collaboration and Continuous Improvement

The dynamic nature of cyber threats requires a collaborative and adaptive approach. FinTech companies must stay informed about the latest threat trends and work closely with industry peers, regulatory bodies, and cybersecurity experts to share insights and develop effective countermeasures. Regular updates to security policies, investments in cutting-edge technologies, and fostering a culture of vigilance across the organization are essential to staying ahead of adversaries.

---

[60] https://fintechmagazine.com/articles/nvidia-advancing-cybersecurity-efforts-with-gen-ai

Conclusion

Cybersecurity threats in FinTech are dynamic, multifaceted, and constantly evolving, reflecting the complexity and interconnected nature of the industry itself. As FinTech continues to revolutionize financial services through innovation and accessibility, it simultaneously becomes a prime target for cybercriminals seeking to exploit vulnerabilities in technology and operations.

To safeguard the trust of users, protect sensitive data, and ensure regulatory compliance, FinTech companies must adopt a proactive and holistic approach to cybersecurity. This includes leveraging cutting-edge technologies, fostering a culture of security awareness, and maintaining robust risk management strategies. Collaboration within the industry, adherence to global best practices, and regular assessments of emerging threats are also crucial to staying ahead in this ever-changing landscape.

By understanding and addressing these risks effectively, FinTech organizations not only protect their customers but also strengthen their own reputations, laying a solid foundation for sustainable growth. In doing so, they contribute to the resilience and advancement of the broader financial ecosystem, ensuring that innovation is matched with security at every step. This balance is critical for the continued trust, adoption, and success of FinTech in a digital-first world.

💡 What We Learnt:

- **Rapid growth in FinTech** has significantly increased exposure to cybersecurity risks.

- **Fraud and identity theft** remain persistent challenges, exploiting stolen data for unauthorized access.

- **Data breaches** are a major concern, often caused by weak encryption or misconfigured systems.

- **Ransomware attacks** disrupt operations by encrypting data and demanding ransom payments.

- **DDoS attacks** overwhelm systems with traffic, leading to downtime and financial losses.

- **Insider threats** pose risks from both malicious and unintentional actions by employees or partners.

- **Phishing attacks** trick users into revealing sensitive information through deceptive messages.

- **Malware threats** compromise devices and systems, leading to data theft or operational disruptions.

- **Social engineering tactics** exploit human error to gain unauthorized access to systems.

- **API vulnerabilities** expose FinTech systems to exploitation by attackers.

- **Weak authentication mechanisms** increase risks of brute-force attacks and credential theft.

- **Unpatched software vulnerabilities** are frequently exploited by cybercriminals.

- **Third-party risks** arise from dependencies on vendors with inadequate security protocols.

- **Robust security measures** like multi-factor authentication and encryption are critical to mitigating threats.

- **User education and awareness** help reduce the impact of phishing and social engineering.

- **Collaboration across the industry** is essential to share insights and improve threat responses.

CHAPTER 4
EMERGING THREATS IN ADVANCED TECHNOLOGIES

In the dynamic world of FinTech, technological innovation continually redefines how financial services are delivered. However, as these technologies evolve, so do the threats associated with them. In this chapter, we delve into emerging risks tied to advanced technologies such as blockchain, artificial intelligence (AI), machine learning (ML), the Internet of Things (IoT), and decentralized finance (DeFi). Each of these innovations presents opportunities but also significant cybersecurity challenges.

📖 Blockchain Security Issues: Safeguarding the Future of Finance

Blockchain technology has revolutionized the financial world with its promise of secure, decentralized systems that offer transparency and autonomy. Despite its transformative potential, the adoption of blockchain has highlighted significant security challenges. Among these, private key management and wallet security are two of the most critical areas requiring urgent attention. Ensuring robust security measures in these domains is essential to fostering user confidence and enabling the safe, widespread adoption of blockchain-based systems.

Private Key Management

At the core of blockchain technology lies cryptographic key pairs, consisting of public and private keys. The private key[61] is particularly critical, functioning as the gatekeeper to blockchain wallets by enabling users to sign transactions and access their digital assets. However, this reliance on private keys introduces a significant vulnerability. A compromised or lost private key can result in the irretrievable loss of funds, as the decentralized nature of blockchain leaves no room for centralized recovery mechanisms.

To address these challenges, several best practices and technologies can be implemented:

Hardware Wallets

Hardware wallets provide an offline, secure solution for storing private keys, significantly reducing the risk of online threats like phishing attacks and malware. By isolating the keys from internet-connected devices, these wallets create an additional security barrier that makes unauthorized access extremely difficult. They are ideal for users who prioritize the safety of their digital assets, especially for long-term storage. Educating users about the benefits of investing in high-quality hardware wallets and understanding their proper usage is crucial to maximizing the protection they offer. Clear guidance on keeping the hardware wallets updated with the latest firmware and avoiding purchasing devices from untrusted sources should also be emphasized.

Multi-Signature Wallets

Multi-signature (or multi-sig) wallets require multiple parties or keys to authorize a single transaction. This setup enhances security by eliminating the risk associated with a single point of failure. Even

---

[61]https://cheatsheetseries.owasp.org/cheatsheets/Key_Management_Cheat_Sheet.html

if one key is compromised, the attacker cannot access the funds without all the required authorizations. This approach is particularly beneficial for organizations and individuals handling significant funds, such as businesses, institutions, or high-net-worth individuals. Implementing multi-sig wallets can deter malicious actors and provide an additional layer of accountability, especially when used in collaborative or enterprise environments. Education on setting up and maintaining multi-sig wallets effectively is essential to ensure they are both secure and user-friendly.

Backup and Recovery Protocols

Secure backup solutions are a cornerstone of cryptocurrency security. Users must create encrypted backups of their private keys and store these backups in multiple, secure physical locations to mitigate the risk of loss due to theft, damage, or natural disasters. Properly safeguarding recovery phrases (seed phrases) is equally critical; these phrases are the ultimate lifeline for regaining access to wallets in case of device failure or loss. Users should store seed phrases offline, in fireproof and waterproof materials, and ensure they are inaccessible to unauthorized individuals. Educating users on best practices for handling backups, such as avoiding digital storage of recovery phrases or sharing them with others, is vital for minimizing risks.

Custodial Solutions for Inexperienced Users

While decentralization is a fundamental principle of blockchain technology, not every user is comfortable managing their private keys independently. For inexperienced users, custodial solutions, where trusted entities manage private keys on their behalf, can serve as an accessible entry point into the crypto ecosystem. These services must adhere to the highest security standards to gain users' trust. Strategies such as employing cold storage for the majority of funds, implementing advanced encryption techniques, and leveraging multi-party computation (MPC) can significantly enhance the security of custodial solutions. Additionally, educating users about the trade-offs between self-custody and custodial solutions ensures they can make informed decisions that align with their risk tolerance and technical expertise.

Wallet Security

Blockchain wallets, as the primary interface for accessing blockchain networks, are a primary target for cybercriminals. Attacks on wallets range from phishing scams and malware to exploiting vulnerabilities in wallet software. Ensuring wallet security is a cornerstone of building trust in blockchain systems and safeguarding user assets.

To enhance wallet security[62], the following measures should be adopted:

End-to-End Encryption

Robust encryption protocols are foundational for safeguarding sensitive transaction data in blockchain ecosystems. Ensuring that all communications between wallets and blockchain networks are encrypted using advanced cryptographic algorithms helps prevent attackers from intercepting, tampering with, or manipulating data. By employing methods such as Transport Layer Security (TLS) or asymmetric encryption, users can be assured that their transactions remain confidential and resistant to interception during transmission. This layer of security is especially crucial in a decentralized system where trust is distributed and external threats are constant.

---

[62] https://www.ssl.com/article/key-management-best-practices-a-practical-guide/

Secure Software Development Practices

Developing wallet software with security as a priority is non-negotiable in today's cyber threat landscape. Developers must adhere to secure software development lifecycle (SDLC) practices, which include rigorous code audits, static and dynamic vulnerability assessments, and penetration testing to identify and fix potential flaws before attackers can exploit them. Incorporating peer reviews into the process ensures additional scrutiny, enhancing code reliability. Open-source wallet projects can also gain significantly from community contributions, as public access to the codebase fosters transparency and increases the chances of discovering and addressing vulnerabilities quickly.

Strong User Authentication Mechanisms

Authentication serves as a critical defense mechanism to prevent unauthorized access to wallets. To ensure robust security, wallet platforms should implement multi-factor authentication (MFA), which requires users to verify their identity through multiple channels, such as something they know (password), something they have (security token), or something they are (biometrics). This multi-layered approach significantly reduces the likelihood of successful breaches, even if a single authentication factor is compromised. Emerging technologies such as biometric authentication—leveraging unique physical traits like fingerprints, facial recognition, or voice patterns—add an extra layer of sophistication to authentication systems, making it nearly impossible for attackers to bypass.

Mitigating Social Engineering Attacks

Social engineering remains one of the most effective tactics employed by cybercriminals to exploit human vulnerabilities. Education and awareness campaigns are essential for equipping users with the knowledge to recognize and avoid common scams, such as phishing emails, fake wallet applications, and malicious websites. Blockchain platforms and wallet providers must prioritize designing user-friendly interfaces and providing clear security guidelines that help users easily identify legitimate platforms. Additionally, integrating features like anti-phishing warnings or alerts for suspicious activity within wallet software can help users stay vigilant.

Continuous Updates and Patches

In the rapidly evolving digital landscape, new vulnerabilities emerge frequently, necessitating continuous updates and patches to wallet software. Developers must establish robust mechanisms for identifying and addressing security flaws promptly. Users should be actively encouraged to keep their wallets updated through automated notifications or reminders that minimize the risk of phishing attempts. Moreover, developers can implement secure update mechanisms that verify the authenticity of software patches before applying them, ensuring that updates themselves do not become a vector for attacks.

Hardware Wallet Integration

Combining software wallets with hardware wallets offers a compelling security solution for users who prioritize protecting their assets. Hardware wallets store private keys in isolated, offline environments, ensuring that these critical pieces of information are never exposed to online threats, even during transaction signing processes. Wallet platforms can provide seamless integration options with hardware wallets to create a hybrid model that enhances usability without compromising security. For high-value transactions, leveraging hardware wallets reduces the risk of theft significantly and ensures an added layer of peace of mind for users.

By adopting these best practices, wallet providers can create robust, user-centric solutions that minimize vulnerabilities and build trust within the blockchain ecosystem.

📖 Building a Resilient Blockchain Ecosystem: Prioritizing Security and Trust

As blockchain technology continues to evolve and gain traction within mainstream financial systems, its potential to revolutionize industries becomes increasingly evident. However, alongside this growth comes a critical responsibility: ensuring the security and integrity of blockchain ecosystems. The decentralized nature of blockchain technology, while its greatest strength, also introduces unique vulnerabilities. It demands that both users and developers adopt proactive measures to protect digital assets, secure transactions, and maintain trust across the network.

The Importance of Security in Blockchain

The core appeal of blockchain lies in its transparency, immutability, and decentralized architecture. These features enable trustless transactions and eliminate the need for intermediaries, paving the way for decentralized finance (DeFi), tokenized assets, and other innovative applications. However, without robust security protocols, these benefits can quickly become liabilities.

Recent years have witnessed a surge in cyberattacks targeting blockchain platforms, from wallet hacks to smart contract exploits and phishing scams. Each incident underscores the importance of establishing and adhering to stringent security practices. To foster a resilient blockchain ecosystem, it is imperative that developers, businesses, and users work together to mitigate these risks.

Proactive Measures for Developers and Users

One of the foundational aspects of blockchain security is private key management. For end-users, private keys are the gateway to their digital assets, and any compromise can result in irreversible losses. To address this, developers must focus on creating user-friendly wallets that offer advanced security features such as multi-signature authorization, biometric authentication, and secure backup options. Simultaneously, users must be educated on the importance of safeguarding their private keys and avoiding common pitfalls such as storing them in insecure locations or sharing them online.

In addition to private key management, the security of smart contracts must be prioritized. Smart contracts, which enable self-executing agreements, are a cornerstone of blockchain innovation. Yet, poorly written or audited smart contracts can introduce vulnerabilities that attackers can exploit. Conducting comprehensive audits, employing formal verification methods, and adopting standardized best practices can significantly reduce these risks.

Collaboration and Innovation

Building a resilient blockchain ecosystem requires collaboration among stakeholders. Developers, industry leaders, academic institutions, and governments must come together to create a cohesive framework for blockchain security. By establishing open standards, sharing threat intelligence, and fostering a culture of continuous improvement, the community can stay ahead of emerging threats.

Education is another critical component. As blockchain adoption grows, so does the need to equip users and developers with the knowledge to navigate this complex landscape safely. Public awareness campaigns, developer training programs, and accessible educational resources can empower individuals and organizations to make informed decisions.

Finally, innovation remains a cornerstone of resilience. Advancements in cryptographic techniques, such as zero-knowledge proofs and homomorphic encryption, hold promise for enhancing privacy and security. Similarly, emerging technologies like decentralized identity solutions and quantum-resistant cryptography can address current and future challenges, ensuring that blockchain ecosystems remain robust and trustworthy.

Safeguarding the Future of Decentralized Finance

By addressing these challenges head-on through a combination of innovation, education, and collaboration, the blockchain community can create a secure and resilient ecosystem. A proactive approach to security will not only safeguard digital assets but also strengthen trust in blockchain technology, encouraging further adoption and investment.

The blockchain revolution holds immense potential to transform global financial systems, supply chains, healthcare, and beyond. However, realizing this potential requires unwavering commitment to security and resilience. By prioritizing these principles today, the blockchain community can build a sustainable foundation for the future, unlocking the full transformative power of decentralized technology.

Risks in AI/ML-Powered FinTech Products: Balancing Innovation and Safety

Artificial intelligence (AI) and machine learning (ML) are revolutionizing the financial technology (FinTech) sector. These technologies automate processes, enhance decision-making, and detect fraudulent activities with unprecedented speed and accuracy. However, with these transformative capabilities come significant challenges and risks. Effectively managing these risks is crucial to ensuring innovation and safety coexist in the rapidly evolving FinTech landscape.

Algorithmic Bias: A Barrier to Fairness and Trust

AI and ML systems learn from historical data, which may inadvertently reflect societal biases. In the FinTech sector, where algorithms[63] drive critical decisions like credit scoring, loan approvals, and insurance underwriting, these biases can have far-reaching implications.

Real-Life Implications

🔍 A notable example is the controversy surrounding an AI-powered credit card program, which allegedly offered women lower credit limits than men[64], even with similar financial profiles. Such disparities can erode trust, attract regulatory scrutiny, and damage an organization's reputation.

Mitigation Strategies

1. **Diverse Training Data**

    FinTech companies must prioritize diversity in training datasets. For example, incorporating data from underserved demographics can help ensure that AI models make equitable decisions. This approach reduces the likelihood of replicating historical biases and promotes inclusivity.

---

[63] https://www.brookings.edu/articles/reducing-bias-in-ai-based-financial-services/
[64] AI-powered credit card program, which allegedly offered women lower credit limits than men

2. **Regular Audits**

Establishing a robust auditing framework is essential. These audits should include fairness tests, stress tests, and independent evaluations. For instance, auditing frameworks similar to Google's AI fairness practices can help organizations identify and mitigate bias in their systems.

3. **Explainable AI**

Enhancing transparency is critical. By developing AI models capable of explaining their decision-making processes, organizations can ensure accountability. For instance, AI explainability can help a rejected loan applicant understand the specific factors that influenced the decision, fostering trust.

AI-Driven Fraud Detection: A Double-Edged Sword

AI and ML have transformed fraud detection[65] by analyzing patterns and identifying anomalies in real-time. However, these systems are not infallible and are increasingly targeted by adversarial AI techniques.

Emerging Threats

1. **Adversarial Attacks**

Attackers can manipulate input data to exploit weaknesses in machine learning models. For example, in one high-profile case, attackers fed fraudulent transaction data designed to mimic legitimate patterns, bypassing fraud detection algorithms[66].

---

[65] https://stxnext.com/blog/ai-in-fintech-threats-risks-and-challenges
[66] https://www.sciencedirect.com/science/article/pii/S016740482300456X

2. **Synthetic Identity Fraud**

   Fraudsters now create synthetic identities—fabricating details that AI systems may not recognize as fraudulent. The rise of synthetic identity fraud has resulted in billions of dollars in losses globally, highlighting the limitations of AI-driven fraud detection when used in isolation.

Mitigation Strategies

1. **Continuous Training**

   Machine learning models must be regularly updated with the latest fraud patterns and data. For example, leading financial institutions like JPMorgan Chase have dedicated AI teams focused on continuously retraining fraud detection models to adapt to new tactics.

2. **Layered Security**

   A hybrid approach that combines AI with traditional methods can offer robust protection. For instance, integrating AI-driven anomaly detection with manual reviews by fraud experts ensures a more comprehensive defense.

3. **Threat Intelligence**

   Leveraging global threat intelligence networks is essential to staying ahead of adversarial techniques. Companies can collaborate with organizations like the Financial Services Information Sharing and Analysis Center (FS-ISAC) to share insights and bolster their defenses.

📖 Operational Risks and Ethical Considerations in AI for FinTech

AI has become a cornerstone of innovation in the FinTech industry, driving efficiencies, automating processes, and enhancing customer experiences. However, this reliance on AI introduces significant operational risks and ethical dilemmas that require careful consideration.

Scalability and Overfitting in AI Models

One critical operational challenge for AI in FinTech is ensuring scalability while avoiding overfitting[67]. Overfitting occurs when a model performs exceptionally well on its training data but struggles to generalize to new datasets. This issue can result in poor decision-making, particularly in dynamic and unforeseen scenarios.

🔍 A stark example of scalability failure occurred in 2021 with a global payment processor. The company experienced significant downtime during the holiday shopping season due to an AI-based fraud detection system[68]. While the system was designed to detect anomalies, it had been overfitted to normal transaction patterns and failed to handle the holiday surge. The result was widespread transaction failures, leaving customers frustrated and merchants unable to process sales. This incident highlights the need for stress-testing AI systems under extreme conditions and designing robust models capable of adapting to unusual scenarios.

---

[67] https://www.financemagnates.com/fintech/ai-risks-in-fintech-10-ai-challenges-fintechs-still-struggle-with/

[68] https://wise.com/gb/blog/ai-for-global-payments

🔍 Another example is in credit risk assessment. AI models trained on historical data may overfit to trends that no longer apply, such as credit behaviors influenced by pre-pandemic conditions[69]. When applied during the COVID-19 pandemic, these models failed to account for abrupt economic changes, leading to poor loan approvals and credit scoring inaccuracies.

📖 Ethical Dilemmas in AI Deployment

AI-driven decisions must align with ethical standards to avoid reputational harm and regulatory consequences. Ethical considerations in FinTech often involve balancing profit motives with fairness, transparency, and consumer protection[70].

Manipulative Marketing Tactics

One prominent ethical dilemma involves AI-powered targeted marketing. For instance, some digital banking platforms have been criticized for leveraging AI to promote impulsive financial behaviors, such as overspending. A notable case involved a popular e-commerce platform that used AI algorithms to deliver hyper-targeted ads encouraging users to buy now and pay later. While effective at driving short-term sales, these tactics contributed to increased consumer debt and drew scrutiny from regulators, who accused the platform of exploiting vulnerable individuals[71].

---

[69] https://www.genpact.com/insight/covid-19-impact-and-recommendations-for-credit-risk-management

[70] https://www.ey.com/en_us/insights/forensic-integrity-services/ai-discrimination-and-bias-in-financial-services

[71] https://www.harvardmagazine.com/2024/03/ai-and-consumerism

Bias in Decision-Making

Bias in AI models also poses ethical risks. In 2019, a major tech company faced backlash after its AI-driven credit card approval algorithm allegedly offered significantly lower credit limits to women compared to men with similar financial profiles. Although the company denied intentional discrimination, the lack of transparency in the AI system's decision-making process fueled public outrage and regulatory investigations[72].

Privacy Concerns

AI systems in FinTech often require vast amounts of personal and financial data, raising concerns about data privacy and misuse. In 2020, a FinTech startup specializing in personalized budgeting tools faced a class-action lawsuit after its AI-powered app was found to have sold user data to third-party advertisers without explicit consent. This breach not only damaged the company's reputation but also highlighted the importance of stringent data governance practices[73].

📖 Mitigation Strategies

To address these challenges, FinTech companies must adopt proactive measures:

1. **Stress-Testing AI Systems**: Companies should simulate extreme conditions, such as transaction surges or market crashes, to ensure their AI systems can scale effectively and generalize to diverse scenarios.

---

[72] https://www.wired.com/story/the-apple-card-didnt-see-genderand-thats-the-problem/
[73] https://www.mdpi.com/2673-4591/32/1/3

2. **Implementing Ethical Guidelines**: Developing clear ethical frameworks for AI deployment is crucial. These guidelines should prioritize transparency, fairness, and consumer welfare over purely profit-driven outcomes.

3. **Regular Audits**: Conducting independent audits of AI models can help identify and mitigate biases, ensuring compliance with regulatory standards.

4. **Transparency in Recommendations**: FinTech firms should provide clear explanations for AI-driven decisions, whether in credit approvals, investment advice, or personalized offers. This transparency builds trust with consumers and regulators alike.

📖 Regulatory Compliance and the Path Forward

With growing regulatory scrutiny, FinTech organizations must ensure their AI/ML systems comply with data privacy laws (e.g., GDPR, CCPA) and financial regulations.

1. **Proactive Engagement with Regulators:** Collaborating with regulators can help companies anticipate and adapt to evolving legal requirements. Initiatives like the UK's Financial Conduct Authority (FCA) sandbox allow FinTech firms to test innovative products in a controlled environment.

2. **Ethics Committees:** Establishing internal ethics committees can provide oversight for AI/ML initiatives, ensuring they align with legal and moral standards.

📖 IoT and DeFi: Expanding the Attack Surface in Financial Services

The Internet of Things (IoT) and decentralized finance (DeFi) are revolutionizing the financial services industry. IoT is streamlining operations with connected devices, while DeFi is democratizing access to financial tools by eliminating intermediaries. However, their rapid adoption introduces unprecedented security challenges, potentially creating vulnerabilities that attackers can exploit. Understanding and mitigating these risks is crucial for financial institutions and technology providers alike.

IoT in Financial Services

The integration of IoT[74] devices, such as connected ATMs, payment terminals, smart card readers, and mobile banking gadgets, is transforming how financial services are delivered. These devices enhance convenience, speed, and personalization for consumers but also expose institutions to a growing array of cyber threats.

Key Concerns in IoT Security

1. **Device Vulnerabilities:** IoT devices often run on lightweight operating systems with limited processing power, which makes implementing strong security measures challenging. In 2021, a cyberattack on a network of IoT payment terminals led to the unauthorized siphoning of customer data, highlighting the vulnerabilities of poorly secured endpoints[75].

2. **Interconnectivity Risks:** The interconnected nature of IoT devices means that a single compromised device can serve

---

[74] https://stxnext.com/blog/ai-in-fintech-threats-risks-and-challenges
[75] https://www.sciencedirect.com/science/article/pii/S2667345223000238

as a gateway for attackers to infiltrate broader systems. For instance, hackers exploited vulnerabilities in a connected thermostat at a casino to steal high-value data from the organization's central network[76].

3. **Data Breaches:** IoT devices collect vast amounts of sensitive data, from transaction histories to geolocation. Without encryption and robust storage mechanisms, this data becomes an attractive target for cybercriminals. A 2022 breach in connected banking kiosks revealed how unprotected data streams could be intercepted and used for fraudulent activities.

Strategies for Securing IoT Ecosystems

To address these risks, financial institutions must adopt a multi-layered approach to IoT security:

- **Firmware Updates:** Ensure IoT devices are updated regularly to fix known vulnerabilities. A connected ATM network in Asia successfully mitigated a series of malware attacks in 2023 by implementing automated firmware update protocols.

- **Network Segmentation:** Isolate IoT networks from critical systems to limit the lateral movement of threats. For example, a global payment processor uses dedicated VLANs for IoT devices, ensuring that breaches in one segment cannot compromise core operations.

- **Device Authentication:** Employ digital certificates and secure protocols like TLS to validate device identities. In 2020, a major US bank integrated hardware-based

---

[76] https://deviceauthority.com/security-and-privacy-issues-in-iot-generated-big-data/

authentication into its IoT payment terminals, significantly reducing unauthorized access attempts[77].

DeFi Platforms

Decentralized finance (DeFi) platforms[78] have disrupted traditional banking by enabling peer-to-peer transactions, lending, and investments without intermediaries. Built on blockchain technology and powered by smart contracts, DeFi is driving financial inclusion and innovation. However, the technology introduces risks that, if unaddressed, could undermine trust in the ecosystem.

Key Threats in DeFi Security

1. **Smart Contract Exploits:** Smart contracts are self-executing codes that automate transactions on DeFi platforms. Bugs or vulnerabilities in these contracts can lead to catastrophic losses. In 2021, the Poly Network hack resulted in $610 million being stolen due to a flaw in the contract logic[79].

2. **Governance Manipulation:** Many DeFi platforms use decentralized governance, where token holders vote on changes. Attackers can accumulate tokens to influence decisions, often to the detriment of the platform. In 2022, a bad actor manipulated the voting system of a prominent DeFi platform, redirecting funds to a private account[80].

3. **Liquidity Pool Risks:** Liquidity pools, where users provide assets for trading, are central to DeFi. However, these pools are vulnerable to flash loan attacks, as seen in the $130

---

[77] https://nvlpubs.nist.gov/nistpubs/SpecialPublications/NIST.SP.1800-16.pdf
[78] https://thecryptocortex.com/key-management-for-blockchain-projects/
[79] https://www.bis.org/publ/othp71.pdf
[80] https://www.bis.org/publ/work1061.pdf

million attack on Cream Finance, where an attacker exploited pricing mechanisms to drain funds[81].

Best Practices for Securing DeFi

- **Formal Verification:** Use mathematical models to verify the correctness of smart contracts. Platforms like Aave and MakerDAO have incorporated formal verification tools to enhance the reliability of their contracts.

- **Insurance Protocols:** Decentralized insurance models can provide users with protection against unforeseen losses. For instance, Nexus Mutual offers coverage for smart contract failures, giving users greater confidence in participating in DeFi.

- **Regular Audits:** Partner with leading security firms to conduct thorough audits of platform code and architecture. Yearn Finance, a prominent DeFi platform, regularly publishes audit results to assure its users of its commitment to security.

📖 The Intersection of IoT and DeFi

The convergence of the Internet of Things (IoT) and Decentralized Finance (DeFi)[82] represents a groundbreaking shift in the technological and economic landscape. By combining IoT's ability to seamlessly connect physical devices with DeFi's trustless and decentralized financial systems, this fusion has the potential to revolutionize industries. One compelling example is the concept of autonomous IoT-enabled payment systems in smart cities. Imagine

---

[81] https://www.aon.com/en/insights/cyber-labs/flash-loan-attacks-a-case-study
[82] https://aws.amazon.com/solutions/guidance/secure-blockchain-key-management-with-aws-nitro-enclaves/

a scenario where connected vehicles automatically pay for tolls, parking fees, or even electric vehicle (EV) charging in real-time using DeFi wallets, eliminating the need for human intervention and reducing transaction friction.

In a broader sense, IoT devices could serve as financial agents capable of executing complex financial transactions. For instance, smart appliances might autonomously negotiate energy contracts with suppliers or micro-sensors embedded in industrial equipment could trigger automated insurance claims using smart contracts. These applications have the potential to streamline operations, reduce costs, and enhance user experience, bringing us closer to a fully automated and interconnected economic ecosystem.

However, the benefits come with significant risks. Integrating IoT with DeFi introduces a hyper-connected ecosystem that expands the potential attack surface exponentially. Vulnerabilities in either IoT devices or DeFi protocols could result in devastating security breaches. Compromised IoT devices might act as entry points for hackers to exploit DeFi platforms, while insecure DeFi smart contracts could provide a pathway to control IoT networks. As these two domains converge, ensuring the security and reliability of the system becomes paramount.

Future Security Considerations

To address the challenges and risks of IoT-DeFi integration, a multi-layered approach to security is essential. Key considerations include:

1. Integration Protocols
Standardizing communication protocols between IoT devices and DeFi platforms is critical to ensuring seamless and secure interactions. Without universal standards, the likelihood of

incompatibilities and vulnerabilities increases. For example, differing encryption methods or inconsistent data validation processes could create exploitable loopholes. A standardized framework would establish baseline security measures, reduce complexity, and enhance interoperability across various devices and platforms[83].

2. Real-Time Monitoring

The dynamic nature of IoT-DeFi transactions necessitates real-time monitoring systems capable of detecting and mitigating threats as they emerge. Leveraging AI-driven analytics can significantly enhance the detection of anomalies and suspicious activities. For instance, a payment network could instantly flag repeated small-value transactions designed to exploit DeFi liquidity pools or identify unusual transaction spikes that deviate from established patterns. By integrating predictive models, such systems could proactively anticipate and block potential attacks before they escalate[84].

3. Cross-Sector Collaboration

The complexity of securing the IoT-DeFi ecosystem calls for collaboration across multiple sectors. IoT manufacturers, DeFi developers, cybersecurity experts, and regulatory bodies must work together to establish industry-wide standards and best practices. For example, partnerships could focus on creating a unified security certification process for IoT devices designed to interact with DeFi platforms. Such efforts would not only enhance security but also build trust among consumers and businesses, accelerating the adoption of these technologies[85].

---

[83] https://aws.amazon.com/blogs/publicsector/4-common-iot-protocols-and-their-security-considerations/

[84] https://datatracker.ietf.org/doc/rfc8576/

[85] https://www.doubloin.com/learn/ethereum-security-and-defi

4. Secure Identity and Access Management

With billions of IoT devices expected to interact with DeFi systems, robust identity and access management protocols are essential. Blockchain-based identity systems could provide a decentralized and tamper-proof method for authenticating devices and users. Additionally, multi-factor authentication and hardware-based security modules could further reduce the risk of unauthorized access[86].

5. Resilient Smart Contracts

Smart contracts form the backbone of DeFi applications. Ensuring their resilience against bugs and malicious exploits is crucial in the context of IoT. Comprehensive code audits, formal verification, and continuous security updates are necessary to maintain the integrity of these contracts. Moreover, implementing mechanisms such as kill switches or governance frameworks could help mitigate damage in the event of an exploit[87].

6. Privacy-Preserving Technologies

The integration of IoT and DeFi involves the exchange of vast amounts of sensitive data, raising significant privacy concerns. Technologies such as zero-knowledge proofs and homomorphic encryption can ensure that transactions and device data remain confidential while still allowing for verification and compliance. These tools could provide a balance between transparency and privacy, fostering trust in the system[88].

---

[86] https://datatracker.ietf.org/doc/rfc8576/
[87] https://www.doubloin.com/learn/ethereum-security-and-defi
[88] https://datatracker.ietf.org/doc/rfc8576/

📖 Conclusion

The integration of blockchain, AI/ML, IoT, and DeFi in financial services marks the dawn of a transformative era, bringing unprecedented innovation and efficiency. These technologies offer immense potential to reshape financial ecosystems, enabling real-time payments, democratized financial access, and enhanced transparency. However, this convergence also presents significant security challenges, including increased attack surfaces and sophisticated cyber threats.

To navigate these challenges, adopting robust security practices is essential. This includes multi-layered defenses, AI-driven threat detection, and privacy-preserving technologies. Cross-sector collaboration among financial institutions, technology providers, regulators, and cybersecurity experts is equally critical to establish unified standards and frameworks.

Finally, continuous education and skill development will ensure the industry stays ahead of evolving threats. By embracing proactive measures and fostering partnerships, the FinTech sector can build a secure, resilient, and innovative future for all.

💡 What We Learnt:

- **Hardware wallets** protect private keys by storing them offline, reducing risks from phishing and malware.
- **Multi-signature wallets** improve security by requiring multiple approvals for transactions.

- **Secure backups** of private keys and recovery phrases are essential to avoid irreversible asset loss.

- **Custodial solutions** provide accessible options for inexperienced users but require strong security practices.

- **End-to-end encryption** safeguards sensitive transaction data from interception.

- **Secure software development practices** mitigate vulnerabilities through rigorous audits and testing.

- **Multi-factor authentication (MFA),** including biometric options, enhances user access security.

- **Continuous updates and patches** effectively address emerging wallet vulnerabilities.

- **Hardware wallet integration** with software wallets creates hybrid security solutions.

- **AI models** can replicate societal biases from training data, affecting fairness in credit and loan decisions.

- **Diverse training datasets**, regular audits, and explainable AI models reduce algorithmic bias.

- **Adversarial attacks and synthetic identity fraud** challenge AI-powered fraud detection systems.

- **Continuous training,** layered security, and threat intelligence improve fraud detection resilience.

- **Overfitting in AI models** can lead to failures during atypical scenarios like transaction surges.

- **Ethical concerns in AI** include manipulative marketing, decision-making biases, and data privacy violations.

- **Stress-testing, ethical guidelines, audits**, and transparent AI decisions address operational and ethical risks.

- **IoT devices in financial services** face vulnerabilities from outdated firmware, interconnectivity risks, and unencrypted data streams.

- **Regular firmware updates, network segmentation**, and device authentication strengthen IoT security.

- **DeFi platforms** are vulnerable to smart contract exploits, governance manipulation, and liquidity pool risks.

- **Formal verification, decentralized insurance, and frequent audits** mitigate DeFi vulnerabilities.

- **IoT-DeFi integration** expands the attack surface but enables applications like autonomous payments.

- **Standardized protocols, real-time monitoring**, and blockchain-based identity systems enhance IoT-DeFi security.

- **Proactive collaboration among stakeholders** is critical for setting standards and sharing threat intelligence.

- **Privacy-preserving technologies** like zero-knowledge proofs and encryption ensure confidentiality and compliance.

- **Continuous education and skill development** are vital for addressing evolving security threats.

CHAPTER 5
UNDERSTANDING AND RESOLVING CYBERSECURITY THREATS

The rapidly evolving landscape of cybersecurity threats presents significant challenges for organizations, particularly in sectors like FinTech that manage sensitive financial data. To stay ahead, businesses must employ comprehensive strategies to identify, understand, and mitigate risks. In this chapter, we explore essential frameworks and methodologies that provide a structured approach to assessing and resolving cybersecurity threats.

📖 Threat Assessment Frameworks

Cybersecurity is not a one-size-fits-all discipline; it requires a tailored approach based on organizational needs, existing vulnerabilities, and potential adversarial behaviors. Effective threat assessment frameworks serve as foundational tools for identifying, analyzing, and mitigating risks.

1. MITRE ATT&CK Framework: Overview and Practical Application

The **MITRE ATT&CK**[89](Adversarial Tactics, Techniques, and Common Knowledge) framework is a widely recognized resource that provides an extensive repository of tactics, techniques, and procedures (TTPs) used by adversaries in the cybersecurity landscape. Designed to help organizations understand, anticipate, and mitigate security threats, the framework serves as a foundational tool for building robust cyber defense strategies.

Overview of the MITRE ATT&CK Framework

Structure:

The framework is structured around adversarial behaviors and is organized into:

- **Tactics**: The "why" behind an adversary's actions, representing the strategic objectives of an attack (e.g., Initial Access, Persistence, or Exfiltration).

- **Techniques**: The "how" these objectives are achieved. Techniques provide a detailed description of the methods attackers use to accomplish their goals.

- **Sub-Techniques**: Granular methods that further break down techniques, providing specificity about adversarial actions.

- **Procedures**: Real-world examples of how specific threat actors implement these techniques.

Purpose:

The primary aim of MITRE ATT&CK is to:

---

[89] https://attack.mitre.org/

- Enable organizations to map real-world adversarial behaviors to their own systems.

- Identify and address security gaps.

- Guide the design of effective security controls and threat detection mechanisms.

By categorizing and mapping threats, the framework fosters a proactive approach to cybersecurity, helping organizations stay ahead of potential adversaries.

Practical Applications in FinTech

FinTech companies are particularly attractive targets for adversaries due to their vast repositories of sensitive financial data and the critical nature of their operations. Leveraging the MITRE ATT&CK framework can significantly enhance the security posture of such organizations in various ways:

1. Gap Analysis

Organizations can compare their existing defenses against the MITRE ATT&CK matrix to pinpoint areas where security controls may be lacking or insufficient. For example:

- **Challenge**: A FinTech company reviews its defenses and discovers limited visibility into insider threats.

- **Action**: By referencing the **"Insider Threat" (TA0001)** tactic, the company identifies techniques like **"Access Token Manipulation" (T1134)** and implements enhanced monitoring tools to mitigate risks.

2. Incident Response Planning

The framework serves as a blueprint for understanding adversarial behaviors, enabling faster containment and recovery during an incident.

🔍 **Example**: A FinTech company experiences a breach involving **phishing**. By referencing the framework, the incident response team quickly identifies that the attack aligns with the **"Initial Access"** tactic and the **"Spear Phishing Attachment" (T1566.001)** technique. The team isolates compromised endpoints, disables affected user accounts, and strengthens email security filters to prevent recurrence.

3. Red and Blue Team Exercises

- **Red Teams** (attack simulations) and **Blue Teams** (defensive teams) use the MITRE ATT&CK framework to simulate and defend against realistic attack scenarios. The matrix provides a common language for both teams, enabling targeted and meaningful exercises.

Scenario:

A FinTech company conducts a simulation focused on a **Credential Dumping (T1003)** attack, where the red team attempts to extract stored user credentials from compromised endpoints. The blue team leverages MITRE ATT&CK to:

- Identify tools the attackers might use (e.g., Mimikatz).

- Implement detection mechanisms for suspicious credential access patterns.

- Practice incident response workflows to improve recovery times.

🔍 Detailed Example of Practical Application

Scenario:

A FinTech company identifies potential risks in its API infrastructure, which underpins critical customer services such as mobile banking, payment gateways, and data aggregation. API misuse could expose the organization to devastating consequences like data breaches or financial fraud.

Step 1: Identifying Threats

The company uses MITRE ATT&CK to analyze potential attack vectors:

- **Tactic**: **Initial Access (TA0001)**

- **Technique**: **Exploitation of Public-Facing Application (T1190)**

- **Sub-Technique**: Attacks such as **API Abuse** (e.g., excessive requests leading to Denial of Service) or **Credential Stuffing** (T1110.004).

Step 2: Mapping Defenses

The team compares these techniques against their current controls. They discover:

- Limited rate-limiting for API requests.

- Insufficient monitoring of failed authentication attempts.

Step 3: Prioritizing Hardening Measures

Based on the findings, the company:

1. Implements rate-limiting to cap the number of requests from a single client within a specific timeframe.

2. Deploys multi-factor authentication (MFA) to protect API endpoints from credential stuffing.

3. Integrates a monitoring tool to log and analyze failed login attempts, flagging potential brute-force attacks.

Step 4: Simulating and Validating Defenses

The red team simulates an API abuse attack, testing the efficacy of rate-limiting and MFA. Concurrently, the blue team monitors logs for suspicious activity, ensuring that detection mechanisms work effectively.

📖 NIST Cybersecurity Framework: Core Principles

The **National Institute of Standards and Technology (NIST)**[90]**Cybersecurity Framework (CSF)** is a widely recognized and adaptable guide designed to help organizations manage and mitigate cybersecurity risks. It provides a structured yet flexible approach, enabling businesses to align cybersecurity activities with their objectives, risk tolerances, and operational needs. The framework is particularly valuable in today's digital landscape, where the frequency and sophistication of cyber threats continue to evolve. Its five core functions—**Identify, Protect, Detect, Respond, and Recover**—serve as foundational pillars for establishing a comprehensive cybersecurity program.

---

[90] https://www.nist.gov/cyberframework

Core Principles of the NIST Cybersecurity Framework

1. Identify

The Identify function emphasizes understanding and managing cybersecurity risks to systems, assets, data, and capabilities. It involves cataloging organizational resources, understanding their value, and assessing vulnerabilities that could be exploited by potential threats.

🔍 **Example**:

A FinTech company identifies and categorizes sensitive customer financial data, including credit card information and account credentials, to determine which data is most critical. They also assess compliance requirements, such as GDPR (General Data Protection Regulation) and PCI DSS (Payment Card Industry Data Security Standard), to ensure alignment with regulatory standards[91].

2. Protect

The Protect function focuses on implementing safeguards to ensure the delivery of critical infrastructure services. It includes deploying technical measures, promoting a culture of cybersecurity awareness, and establishing robust access control mechanisms.

🔍 **Example**:

A FinTech organization deploys multi-factor authentication (MFA) to secure customer accounts and encrypts sensitive data to protect it during storage and transmission. Regular employee training on phishing awareness is also part of the protection strategy[92].

---

[91] https://www.nist.gov/cyberframework/identify
[92] https://www.infosecinstitute.com/resources/nist-csf/nist-csf-core-functions-identify/

3. Detect

Detect involves establishing capabilities to quickly identify cybersecurity incidents, minimizing the time an attacker remains undetected. This can include implementing advanced monitoring systems and regularly updating detection tools to stay ahead of evolving threats.

🔍 **Example**:

A FinTech platform uses artificial intelligence-driven anomaly detection systems to monitor transactional behavior in real time. These systems flag unusual activities, such as multiple failed login attempts or transfers exceeding normal thresholds, enabling swift investigation[93].

4. Respond

The Respond function is centered on taking action to contain the impact of a cybersecurity event. It involves creating detailed incident response plans, conducting regular simulations, and ensuring communication channels are ready to manage incidents effectively.

🔍 **Example**:

In response to a ransomware attack, a FinTech organization executes its pre-defined incident response plan. This includes isolating affected systems, notifying stakeholders, and engaging cybersecurity specialists to contain the attack and begin recovery[94].

---

[93] https://www.infosecinstitute.com/resources/nist-csf/nist-csf-core-functions-identify/
[94] https://www.compassitc.com/blog/the-nist-cybersecurity-framework-the-identify-function

5. Recover

Recover emphasizes the importance of resilience and the ability to restore normal operations after a cybersecurity incident. It includes planning for disaster recovery, learning from incidents, and continuously improving security measures.

🔍 **Example**:

A FinTech company tests its disaster recovery protocols quarterly to ensure they can minimize downtime after an attack. These tests include restoring data from backups, validating the functionality of systems, and evaluating response times[95].

Function	Key Activities	Real-Life Example
Identify	Asset inventory, risk assessment, regulatory mapping	A FinTech company identifies critical customer data and maps it to PCI DSS and GDPR requirements.
Protect	Access control, data encryption, employee training	A company implements MFA, encrypts data with AES-256, and trains employees on phishing awareness.
Detect	Anomaly detection, monitoring, updating detection tools	An AI-driven system flags unusual activity, like rapid login attempts from multiple IPs.
Respond	Incident response plans, stakeholder communication	A cryptocurrency exchange isolates affected servers during a ransomware attack and engages experts for mitigation.
Recover	Data restoration, process improvement	A healthcare provider restores systems using backups, validates systems for malware, and invests in a SIEM system for enhanced monitoring.

---

[95] https://www.device42.com/compliance-standards/nist-csf-categories/

📖 Relevance of the NIST Cybersecurity Framework to FinTech

Adapting to a Dynamic Threat Landscape

The FinTech sector is uniquely positioned at the intersection of technological innovation and financial services. With rapid digitalization, FinTech organizations handle vast transaction volumes, sensitive customer data, and often operate in a globalized environment. This makes them prime targets for cybercriminals seeking financial gain, intellectual property theft, or even large-scale fraud.

🔍 For instance, high-profile breaches like the **Equifax data breach in 2017**[96] highlighted the devastating consequences of inadequate cybersecurity measures. In the FinTech space, cybercriminals have increasingly targeted APIs, digital wallets, and blockchain platforms, exploiting vulnerabilities to siphon funds or compromise customer data[97].

Adopting the **NIST Cybersecurity Framework (CSF)** allows FinTech organizations to keep pace with this ever-changing threat landscape. By emphasizing adaptive risk management practices, the framework ensures companies remain resilient against sophisticated cyber threats such as ransomware, phishing, and zero-day vulnerabilities.

🔍 For example, a FinTech company using blockchain for secure transactions can integrate NIST CSF guidelines to

---

[96] https://www.csoonline.com/article/567833/equifax-data-breach-faq-what-happened-who-was-affected-what-was-the-impact.html
[97] https://blog.rsisecurity.com/top-information-security-frameworks-for-fintech/

harden its systems against **51% attacks**, where a malicious actor gains control over more than half of the blockchain's mining power[98].

Tailoring to Operational Needs

One of the NIST CSF's greatest advantages is its **customizability**, allowing organizations to address their unique operational needs without adopting a "one-size-fits-all" approach. For FinTech companies, this means tailoring the framework to challenges like:

- **Managing Large-Scale Digital Wallets**: Digital wallets, such as those provided by **PayPal** or **Venmo**, process millions of transactions daily. By adopting the CSF, these platforms can ensure end-to-end encryption, robust authentication protocols, and continuous monitoring to detect anomalies in real time.

- **Integrating Blockchain Technologies**: Platforms like **Ripple** and **Ethereum**, which facilitate cryptocurrency transactions, can use the CSF to enhance the security of smart contracts. By addressing risks such as code vulnerabilities or private key management, FinTech firms can prevent multi-million-dollar breaches like the **DAO hack of 2016**[99].

- **Securing API Ecosystems**: Many FinTech firms rely on APIs for third-party integrations, such as connecting with banks, payment processors, or identity verification services. NIST CSF guidelines can help secure these APIs against

---

[98] https://fcicyber.com/top-5-ways-the-financial-services-industry-can-leverage-nist-for-cybersecurity-compliance/
[99] https://www.imf.org/-/media/Files/Publications/WP/2018/wp18143.ashx

threats like **man-in-the-middle attacks** or **data injection exploits**.

Through its flexibility, the NIST CSF allows FinTech companies to adopt best practices suited to their specific technology stack and operational model, ensuring comprehensive risk mitigation.

Regulatory Compliance

FinTech companies operate in a heavily regulated environment. From global standards like **GDPR** (General Data Protection Regulation) for data privacy in the EU to payment security regulations like **PCI DSS** (Payment Card Industry Data Security Standard), FinTech organizations must juggle compliance with various frameworks. Additionally, many countries, including the U.S., impose local regulations like the **California Consumer Privacy Act (CCPA)** or the **New York Department of Financial Services Cybersecurity Regulation** (NYDFS).

The NIST CSF serves as a **bridge between regulatory frameworks** and operational security measures. Its alignment with various standards ensures that FinTech firms can maintain compliance while improving their overall cybersecurity posture.

🔍 For example:

- A payment processing platform like **Stripe** could leverage the NIST CSF to meet PCI DSS requirements by implementing robust encryption standards, regular vulnerability assessments, and continuous monitoring of payment systems.

- Similarly, a FinTech company operating in the EU, such as **Revolut[100]**, can use the NIST CSF to align its data protection practices with GDPR requirements, ensuring customer data is stored, processed, and shared securely.

The framework's mapping capabilities streamline compliance audits, reduce administrative burdens, and allow organizations to focus resources on proactive security measures rather than reactive firefighting[101].

Fostering Trust and Resilience

Trust is fundamental in the FinTech industry. Customers entrust their financial assets and sensitive data to these platforms, expecting reliability and robust security measures. A single data breach can erode trust and have long-term implications for customer loyalty and brand reputation.

The **NIST CSF's core functions—Identify, Protect, Detect, Respond, and Recover—** help FinTech companies build a cybersecurity posture that inspires confidence among stakeholders. For instance:

- A company like **Robinhood[102]**, which offers stock trading services, can showcase its commitment to cybersecurity by implementing advanced detection systems based on the CSF guidelines. This builds investor confidence, particularly after incidents like the **2021 Robinhood data breach**, where unauthorized access impacted millions of users.

---

[100] https://blog.pcisecuritystandards.org/nist-mapping
[101] https://www.standardfusion.com/blog/mapping-pci-dss-to-nist-csf
[102] https://newsroom.aboutrobinhood.com/robinhood-announces-data-security-incident-update/

- **Zelle**, a popular peer-to-peer payment service, could utilize the framework's Protect function to implement strong multifactor authentication (MFA) and encryption protocols, thereby reinforcing its reputation as a secure platform[103].

By leveraging the CSF, FinTech companies can demonstrate their commitment to safeguarding customer data and ensuring uninterrupted service, fostering trust among customers, partners, and regulators.

Strengthening Incident Response and Recovery

In the fast-paced world of financial transactions, any disruption—whether caused by a cyberattack or system failure—can have significant repercussions. For FinTech firms, the ability to quickly respond to and recover from incidents is critical to minimizing both financial and reputational damage.

The **Respond and Recover functions** of the NIST CSF emphasize the importance of having well-documented incident response plans, regular tabletop exercises, and robust recovery protocols. These practices ensure that FinTech companies can limit the impact of a breach and resume normal operations quickly.

🔍 For example:

- When **Capital One** experienced a major data breach in 2019, the company faced criticism for delays in detecting and responding to the attack. FinTech firms can avoid such situations by integrating CSF guidelines, which prioritize

---

[103] https://newsroom.aboutrobinhood.com/robinhood-announces-data-security-incident-update/

real-time detection and streamlined communication during incidents[104].

- In cases of ransomware attacks, platforms like **Chime** or **Cash App** could use the CSF to develop playbooks for isolating infected systems, communicating with affected stakeholders, and restoring operations without succumbing to ransom demands.

By focusing on resilience, the NIST CSF ensures that FinTech companies are prepared to navigate the high stakes of cybersecurity incidents, safeguarding their financial stability and customer trust.

📖 Threat Modeling: Evaluating Risks During Product Development

In today's fast-paced technological landscape, integrating security into the product development lifecycle is no longer optional; it is a necessity. **Threat modeling** provides a structured approach to proactively identify, understand, and address potential security risks during the design and development phases. By systematically analyzing potential attack vectors, organizations can preemptively mitigate vulnerabilities, safeguarding their products and users before exploitation becomes a reality.

📖 What is Threat Modeling?

At its core, threat modeling is a proactive and iterative process aimed at uncovering security weaknesses and addressing them early. It is not a one-size-fits-all process but rather a tailored strategy that

---

[104] https://cams.mit.edu/wp-content/uploads/capitalonedatapaper.pdf

evolves alongside the product. It involves understanding assets, anticipating potential threats, and implementing solutions in alignment with the product's architecture and functionality.

For industries like FinTech, where sensitive financial data and high stakes are the norm, threat modeling becomes even more critical. A minor vulnerability could lead to catastrophic financial losses, reputational damage, and regulatory penalties.

Key Steps in Threat Modeling

1. Identify Assets and Entry Points

The first step in threat modeling is to understand what requires protection and how adversaries might gain access. This includes:

- **Identifying critical assets:** Assets can range from sensitive customer data (e.g., personally identifiable information and financial records) to operational systems like payment gateways and databases.

- **Mapping access points:** Pinpointing interfaces where attackers could potentially exploit vulnerabilities, such as APIs, user authentication portals, mobile applications, or third-party integrations.

  🔍 **Example:** A FinTech company building a mobile banking application identifies customer account data, transactional data, and the payment processing system as critical assets. Entry points include mobile app logins, API integrations for payment processors, and the backend database[105].

---

[105] https://owasp.org/www-community/Threat_Modeling_Process

2. Assess Threat Scenarios

Once assets and access points are identified, the next step is to evaluate how these might be exploited by adversaries. This includes:

- **Mapping potential attack vectors:** For instance, an attacker could exploit weak authentication mechanisms, outdated software, or unsecured APIs.

- **Developing threat scenarios:** Predicting real-world attacks helps create actionable insights. For example, attackers might deploy phishing campaigns targeting employees with access to the database or use a brute-force attack on a poorly implemented login system.

🔍 **Example:** In the mobile banking application scenario, the team predicts possible session hijacking attacks where attackers could intercept session tokens to gain unauthorized access. Another threat scenario involves a Distributed Denial of Service (DDoS) attack targeting payment processing APIs[106].

3. Prioritize Mitigations

With a list of potential threats, the next step is to prioritize mitigation efforts based on:

- **Likelihood of occurrence:** Assessing how easily an attacker could exploit a vulnerability[107].

- **Potential impact:** Evaluating the consequences of an exploited vulnerability, such as data breaches, financial loss, or reputational harm.

---

[106] https://www.exabeam.com/blog/infosec-trends/top-8-threat-modeling-methodologies-and-techniques/
[107] https://blog.techprognosis.com/prioritizing-risk-mitigation-based-on-likelihood-and-impact/

- **Cost-effectiveness:** Allocating resources to address high-risk vulnerabilities in a feasible and efficient manner.

🔍 **Example:** The FinTech company ranks session hijacking as a high-priority risk due to its potential to compromise customer accounts. A low-priority risk might involve cosmetic glitches in the app that have minimal security implications.

4. Test and Validate

Threat modeling is not a one-time exercise; it requires ongoing evaluation to ensure that implemented mitigations remain effective as the threat landscape evolves. Regular testing and updates to the threat model are crucial.

- **Penetration testing:** Simulating attacks to validate the strength of implemented security measures.

- **Feedback loops:** Incorporating insights from emerging threats, user feedback, and regulatory updates to refine the model.

🔍 **Example:** The mobile banking app undergoes penetration testing to simulate a session hijacking attempt. The team validates that the implemented measures—such as secure session token handling and multi-factor authentication (MFA)—effectively mitigate the risk[108].

---

[108] https://www.6sigma.us/six-sigma-in-focus/risk-prioritization-matrix/

📖 Benefits of Threat Modeling in FinTech

Security by Design

Embedding threat modeling into the development lifecycle ensures that security is an integral part of the product's foundation rather than being treated as an afterthought. This approach allows potential vulnerabilities to be identified and mitigated during the design phase, reducing the risk of introducing exploitable flaws into the final product. By addressing security concerns early, FinTech companies can create robust systems that are less susceptible to breaches, which in turn enhances customer trust. Additionally, integrating security into the design phase minimizes the need for costly, time-consuming post-launch fixes and ensures a more seamless user experience.

Cost Savings

Addressing vulnerabilities during development is far more cost-effective than remediating them after a security incident or breach. Threat modeling allows FinTech organizations to identify potential threats and implement appropriate controls before they manifest into significant issues. Early detection and mitigation reduce the need for expensive system patches, legal settlements, compensating affected customers, or paying regulatory fines. Moreover, by avoiding breaches, organizations can prevent reputational damage that often leads to customer attrition, further contributing to long-term financial savings. Proactive security planning through threat modeling thus becomes an investment that protects both the company's resources and its bottom line.

Regulatory Compliance

The FinTech industry operates within a highly regulated environment, with frameworks like GDPR, PCI DSS, and PSD2 requiring organizations to maintain robust security measures to protect sensitive customer data. Threat modeling plays a critical role in achieving compliance by identifying risks and ensuring they are addressed through appropriate controls. Furthermore, the documentation generated during the threat modeling process serves as evidence of due diligence during regulatory audits. Proactively aligning with compliance standards not only avoids fines and penalties but also establishes the organization as a trusted and responsible entity in the eyes of both regulators and customers. This can lead to competitive advantages and increased market credibility.

By leveraging threat modeling, FinTech companies can build secure, cost-efficient, and compliant systems that protect customer data while maintaining business integrity.

📖 Case Study: Mitigating Session Hijacking in a FinTech Startup

Scenario:

A FinTech startup developing a mobile payment app wanted to ensure robust security before launching its product[109]. During the threat modeling process, the team identified session hijacking as a critical vulnerability. In this attack, an adversary intercepts session tokens—unique identifiers used to authenticate a user's ongoing session—to impersonate the user and gain unauthorized access.

---

[109] https://www.pingidentity.com/en/resources/blog/post/session-hijacking.html

Threat Assessment:

The team mapped out the attack scenario:

- A malicious actor intercepts session tokens transmitted over an insecure network.

- Using these tokens, the attacker accesses customer accounts, potentially initiating unauthorized transactions.

Mitigation Strategy:

To address this, the team implemented:

1. **End-to-End Encryption (E2EE):** Ensuring all communications, including session tokens, are encrypted[110].

2. **Token Expiry and Regeneration:** Limiting the validity of session tokens and regenerating them at regular intervals to reduce exposure time.

3. **Multi-Factor Authentication (MFA):** Adding a second layer of authentication to prevent unauthorized access, even if session tokens are compromised.

Validation:

The team conducted penetration testing by simulating session hijacking attacks. Tests revealed that the encryption and token management mechanisms effectively thwarted these attempts. The app also underwent independent security audits to validate its defenses.

Outcome:

By addressing session hijacking pre-launch, the startup avoided a potentially catastrophic breach. Not only did this safeguard

---

[110] https://www.amilma.digital/blog/unmasking-session-hijacking-insights-into-understanding-detecting-and-preventing-this-cyber-threat

customer trust, but it also reinforced the company's reputation as a security-conscious FinTech provider.

📖 Cybersecurity Threat Analysis

Effective cybersecurity threat analysis[111] is essential for organizations, especially those in the FinTech sector, where sensitive data and critical operations are prime targets. This guide explores the key questions driving a robust threat analysis and dives into identifying vulnerabilities within FinTech products, using real-world examples and actionable insights[112].

Key Questions to Drive a Thorough Threat Analysis

To safeguard critical assets, an organization must approach cybersecurity with clarity and focus. The following questions form the backbone of a strategic threat analysis:

1. What assets are at risk?

Every organization possesses assets that are vital to its operations and therefore appealing to malicious actors. These assets can be categorized as follows:

- **Physical Assets:** These include hardware such as servers, network equipment, and endpoint devices.

  🔍 *Example:* A central server hosting transaction records for a digital wallet provider becomes a prime target for ransomware attacks[113].

---

[111] https://www.isaca.org/resources/isaca-journal/issues/2022/volume-5/addressing-the-complexities-of-cybersecurity-at-fintech-enterprises

[112] https://masterofcode.com/blog/generative-ai-for-fintech

[113] https://edge.arista.com/untangle/the-role-of-cryptocurrency-in-ransomware-attacks/

- **Digital Assets:** Software platforms, APIs, and cloud-based services used to deliver FinTech solutions are highly susceptible to breaches.

  🔍 *Example:* A poorly configured API used in mobile banking applications could allow unauthorized data access[114].

- **Human Resources:** Employees and third-party vendors with access to sensitive systems are often targeted via phishing or social engineering.

  🔍 *Example:* A customer service representative's credentials are phished, allowing attackers to access internal systems[115].

- **Intellectual Property:** Proprietary algorithms and systems designed for fraud detection or credit scoring can be exploited if compromised.

  🔍 *Example:* A competitor hacks into a FinTech startup's backend to steal and replicate its predictive analytics model[116].

What data is sensitive?

For FinTech companies, sensitive data often includes:

- **Financial Information:** This encompasses bank account details, credit card numbers, and transaction histories.

---

[114] https://lisnr.com/resources/blog/securing-digital-wallets-addressing-top-security-concerns-for-financial-services-leaders/
[115] https://edge.arista.com/untangle/the-role-of-cryptocurrency-in-ransomware-attacks/
[116] https://lisnr.com/resources/blog/securing-digital-wallets-addressing-top-security-concerns-for-financial-services-leaders/

🔍 *Example:* A breach at a payment processing company leads to exposure of encrypted credit card data, raising concerns about PCI DSS compliance[117].

- **Personally Identifiable Information (PII):** Names, addresses, social security numbers, and other identifiers are invaluable to attackers.

🔍 *Example:* A data breach at a loan provider exposes PII, resulting in identity theft for thousands of customers[118].

- **Regulatory Data:** Compliance-related data, such as records maintained for Anti-Money Laundering (AML) and Know Your Customer (KYC) protocols, must be protected to avoid legal repercussions.

🔍 *Example:* An unsecured database storing KYC documents is discovered by a security researcher, leading to public and regulatory scrutiny[119].

How to Identify the Most Critical Vulnerabilities in FinTech Products

Understanding where vulnerabilities exist is critical to preempting attacks. FinTech products, by their nature, have specific weak points that attackers may exploit. Below are strategies to identify these vulnerabilities, with practical examples:

---

[117] https://etactics.com/blog/pci-violations-and-consequences
[118] https://etactics.com/blog/pci-violations-and-consequences
[119]https://www.pcisecuritystandards.org/documents/Responding_to_a_Cardholder_Data_Breach.pdf

1. Mobile Applications

Mobile applications are central to the FinTech ecosystem but often face risks such as insecure code, insufficient encryption, or poor session management.

- **Threat Vector:** Mobile malware can exploit hardcoded credentials or unencrypted sensitive data stored locally.

  🔍 *Example:* A mobile wallet app stores user PINs locally in plaintext, making them accessible to malware installed on the device[120].

- **Mitigation:** Conduct static and dynamic code analysis, and ensure that all sensitive data is encrypted using industry standards.

2. APIs

APIs enable integrations across platforms but are also prime targets due to insufficient authentication or overly permissive access controls.

- **Threat Vector:** An attacker uses a broken object-level authorization vulnerability in an API to access other users' transaction histories.

  🔍 *Example:* A FinTech API inadvertently allows attackers to retrieve account details by altering the account ID in API requests.

- **Mitigation:** Perform penetration testing focused on API endpoints and adopt least-privilege access controls.

---

[120] https://www.security.com/threat-intelligence/exposing-danger-within-hardcoded-cloud-credentials-popular-mobile-apps

3. Payment Gateways

Payment gateways process financial transactions and are common targets for attackers aiming to intercept or manipulate data.

- **Threat Vector:** A man-in-the-middle attack captures unencrypted card details during payment processing.

  🔍 *Example:* An e-commerce site's payment gateway doesn't enforce HTTPS, allowing attackers to harvest payment information[121].

- **Mitigation:** Use Transport Layer Security (TLS) for all communications and implement tokenization to replace sensitive data during transactions.

4. Cloud Infrastructure

Many FinTech companies leverage cloud services, but misconfigurations in these environments can expose sensitive data.

- **Threat Vector:** Publicly accessible cloud storage buckets containing sensitive customer records.

  🔍 *Example:* A FinTech startup's AWS S3 bucket is left open, exposing transaction logs that include user details.

- **Mitigation:** Use cloud configuration monitoring tools to identify and correct risky settings, and implement encryption for data at rest.

5. Third-Party Integrations

Third-party providers often have access to systems and data, creating supply chain risks.

---

[121] https://www.cm-alliance.com/cybersecurity-blog/payment-gateways-key-element-of-cybersecurity-in-online-transactions

- **Threat Vector:** A vulnerability in a third-party vendor's software becomes an entry point for attackers into a FinTech company's network.

  🔍 *Example:* An accounting software used by a FinTech firm is breached, allowing lateral movement into sensitive systems[122].

- **Mitigation:** Regularly audit third-party software for vulnerabilities and limit access to essential systems only.

📖 Mitigation Strategies and Resolutions for Common Threats[123]

1. Preventing Fraud and Identity Theft

Fraud and identity theft remain significant challenges in the digital age, driven by the increasing sophistication of cybercriminals. Organizations and individuals can adopt a multi-layered approach to prevent these threats:

- **Awareness and Training:** Educating individuals and employees about phishing scams, fake websites, and other social engineering tactics reduces susceptibility.

- **Secure Transactions:** Use encrypted communication channels (e.g., HTTPS) for online transactions. Avoid conducting sensitive operations over public or unsecured Wi-Fi networks.

---

[122] https://www.exabeam.com/explainers/information-security/software-supply-chain-attacks-attack-vectors-examples-and-6-defensive-measures/
[123] https://blog.rsisecurity.com/top-information-security-frameworks-for-fintech/

- **Strong Password Policies:** Encourage the use of complex, unique passwords and periodic changes. Password managers can help maintain security without compromising convenience.

- **Regular Monitoring:** Promptly detect unauthorized transactions or account access by monitoring financial statements and credit reports.

2. Role of Multifactor Authentication (MFA) and Biometrics in Reducing Identity Theft

Multifactor Authentication (MFA) and biometric verification have revolutionized how identity theft risks are mitigated. These technologies ensure that even if one layer of security is breached; additional barriers prevent unauthorized access.

MFA Implementation:

- Combines "something you know" (passwords), "something you have" (security tokens or mobile devices), and "something you are" (biometrics).

- 🔍Examples include SMS-based codes, authenticator apps, or physical security keys[124].

- MFA significantly reduces the risk of account breaches, even if credentials are compromised.

Biometric Authentication:

- Biometric features such as fingerprints, facial recognition, and voice recognition provide a robust identity validation mechanism.

---

[124] https://www.daito.io/resources/2fa-vs-otp/

- Harder to replicate than passwords, biometrics enhance security without relying solely on user memory.

- However, organizations must ensure biometric data is stored securely using encryption and cannot be reverse-engineered.

3. Resolving Insider Threat Risks Through Enhanced Access Controls and Monitoring

Insider threats, whether intentional or accidental, can cause substantial damage. To address these risks, organizations should prioritize stringent access controls and continuous monitoring:

- **Principle of Least Privilege (PoLP):** Grant employees access only to the data and systems they need for their roles. Regular audits can ensure privileges are not excessive or outdated.

- **Access Monitoring:** Track and log user activities to identify unusual behavior. Tools like User and Entity Behavior Analytics (UEBA) can flag anomalies such as unauthorized file access or data transfers.

- **Segregation of Duties:** Divide critical functions across multiple individuals to minimize the risk of fraud or data misuse by a single employee.

- **Awareness Programs:** Train employees to recognize security policies and report suspicious activities.

- **Incident Response Plans:** Develop clear protocols to respond to suspected insider threats swiftly and effectively.

4. Best Practices for Securing Digital Wallets

Digital wallets, widely used for financial transactions and cryptocurrency management, are lucrative targets for cybercriminals. Adopting comprehensive security measures can protect these wallets from theft or unauthorized access:

- **Encryption:** Employ end-to-end encryption for wallet transactions and data storage. Use wallets that comply with recognized security standards.

- **Private Key Management:** Store private keys in secure, offline environments like hardware wallets or encrypted USB drives. Avoid sharing keys or storing them in plaintext on digital devices.

- **Two-Factor Authentication (2FA):** Enable 2FA for wallet access to add an additional layer of protection.

- **Backup Strategies:** Regularly back up wallet data and private keys in secure, offline locations. Use multiple backups stored in geographically dispersed locations for added safety.

- **Software Updates:** Keep wallet applications and associated software updated to protect against vulnerabilities.

- **Be Wary of Phishing:** Only use official wallet applications and verify URLs before entering credentials. Avoid unsolicited links claiming wallet updates or issues.

By implementing these strategies and best practices, individuals and organizations can reduce the risks of fraud, identity theft, insider threats, and digital wallet compromises, fostering a safer digital ecosystem.

📖 Tools and Techniques for Proactive Threat Resolution

Proactive threat resolution is a critical aspect of modern cybersecurity strategies. By addressing potential security threats before they materialize, organizations can minimize risks and ensure the continuity of their operations. Below, we delve into three primary categories of tools and techniques for proactive threat resolution, providing detailed insights and examples.

1. Incident Response and Management Tools

Incident response and management tools are pivotal in addressing security breaches swiftly and effectively. These tools help organizations establish a structured approach to identifying, containing, eradicating, and recovering from threats.

Key Features and Capabilities:

- **Centralized Incident Tracking:** These tools provide a centralized platform for logging and monitoring all incidents, ensuring that no threat is overlooked.

- **Collaboration and Communication:** They facilitate real-time communication between stakeholders, enabling coordinated responses.

- **Playbook Automation:** Many incident management tools come with pre-configured playbooks that automate routine response tasks, reducing response time and human error.

- **Post-Incident Analysis:** Tools generate reports and insights post-resolution to identify root causes and improve future responses.

🔍 **Examples:**

- **Splunk Phantom:** Offers robust orchestration and automation for incident response processes[125].

- **IBM Resilient:** Known for its dynamic playbooks and integration with various security tools.

- **ServiceNow Security Operations:** Helps streamline incident workflows and provides actionable insights through machine learning.

🔍 **Use Case Example:** Imagine a financial institution detecting unusual login attempts. Using an incident response tool like IBM Resilient, the security team could automatically isolate affected systems, notify stakeholders, and initiate an investigation while simultaneously deploying containment measures based on predefined playbooks[126].

2. *Automated Detection and Response*

Automated detection and response (ADR) tools represent a game-changer in cybersecurity. These tools leverage advanced technologies such as machine learning and artificial intelligence to identify threats and respond autonomously, often before human teams are even aware of the issue.

Key Features and Capabilities:

---

[125] https://www.splunk.com/en_us/products/splunk-security-orchestration-and-automation.html
[126] https://www.threatintelligence.com/blog/incident-response

- **Real-Time Threat Detection:** Continuous monitoring ensures threats are identified the moment they arise.

- **AI and ML Integration:** These tools learn from historical data to predict and prevent potential attacks.

- **Automated Response Actions:** ADR tools can isolate affected endpoints, block malicious IPs, or enforce security policies without manual intervention.

- **Scalability:** They handle vast amounts of data, making them ideal for organizations with large and complex IT infrastructures.

⚲ Examples:

- **CrowdStrike Falcon:** A leading endpoint detection and response tool that uses AI to detect and neutralize threats in real time.

- **Microsoft Defender for Endpoint:** Provides advanced threat protection and automated remediation across diverse environments.

- **Palo Alto Cortex XDR:** Offers extended detection and response capabilities, integrating network, endpoint, and cloud data.

⚲ **Use Case Example:** A retail company experiences a sudden surge of phishing attempts targeting its employees. An ADR tool like CrowdStrike Falcon could immediately flag suspicious emails, quarantine affected systems, and update email filters to block similar future threats—all without human intervention[127].

3. Incorporating External Threat Intelligence Feeds and Information Sharing

In today's interconnected world, external threat intelligence and collaborative information sharing are indispensable for proactive threat resolution. By tapping into global threat data, organizations can gain insights into emerging attack vectors and adopt defensive measures accordingly.

Key Features and Capabilities:

- **Real-Time Intelligence Feeds:** Provide up-to-date information on the latest malware, vulnerabilities, and adversary tactics.

- **Threat Indicators:** Include IP addresses, domain names, file hashes, and other identifiers linked to malicious activities.

- **Collaborative Platforms:** Facilitate information sharing within industry groups, fostering a collective defense approach.

- **Integration with Security Tools:** External feeds can be directly integrated into SIEMs (Security Information and Event Management systems) and other tools for actionable insights.

---

[127] https://www.crowdstrike.com/en-us/blog/falcon-sensor-issue-use-to-target-crowdstrike-customers/

🔍 Examples:

- **ThreatConnect:** Combines threat intelligence aggregation with collaborative features.

- **Recorded Future:** Offers predictive threat intelligence by analyzing billions of data points daily.

- **Information Sharing and Analysis Centers (ISACs):** Industry-specific groups (e.g., FS-ISAC for financial services) that promote the exchange of threat intelligence.

🔍 **Use Case Example:** A healthcare organization partners with its sector's ISAC[128] and integrates Recorded Future's threat intelligence feed into its SIEM. When a new ransomware strain targeting healthcare systems is detected globally, the organization receives alerts in real-time and implements preventive measures, such as patching vulnerabilities and enhancing firewall rules.

📖 Post-Threat Analysis and Continuous Improvement

In today's rapidly evolving FinTech ecosystem, threats—whether they are technical, operational, or external—are a constant reality. Successfully addressing these threats is essential for business continuity, but the journey doesn't end with resolution. Post-threat analysis and the adoption of a culture of continuous improvement are key to building resilience and ensuring proactive defense within FinTech product teams. This document delves into the steps to take

---

[128] https://health-isac.org/

post-threat, the mindset needed for resilience, and lessons from practical examples and case studies[129].

Steps to Take After a Threat Has Been Resolved

A resolved threat provides a golden opportunity to learn, improve, and strengthen defenses. The following steps can help ensure that the organization not only resolves the immediate concern but also minimizes future vulnerabilities.

a. Gathering Insights and Incident Documentation

- **Immediate Incident Debrief**: Conduct a post-mortem meeting with key stakeholders to capture what happened, how it was handled, and outcomes. Include a timeline of events, actions taken, and communication logs.

- **Data Analysis**: Analyze logs, monitoring data, and system performance metrics to identify anomalies or weak points. For example, in a Distributed Denial of Service (DDoS) attack, assess network traffic patterns to pinpoint when and how the breach occurred.

- **Stakeholder Feedback**: Gather input from involved team members to understand challenges faced during the response and recovery phases. This input highlights areas needing procedural improvements.

b. Identifying Root Causes

- **Technical Root Cause Analysis (RCA)**: Use RCA frameworks like the 5 Whys or Fishbone Diagrams to trace the threat back to its origin. For instance, a security

---

[129] https://www.americanbanker.com/news/nist-updates-cybersecurity-framework-emphasizing-governance

vulnerability might arise from outdated libraries or misconfigurations.

- **Process and Human Factors**: Assess whether procedural gaps or human errors contributed to the threat. For example, if phishing led to a security compromise, was the training program insufficient, or were technical controls like email filters inadequate?

- **External Influences**: Evaluate external factors such as vendor dependencies or evolving regulatory landscapes that may have contributed to the incident.

c. Updating Response Plans

- **Refinement of Playbooks**: Incorporate lessons learned into incident response playbooks, ensuring they cover the specifics of the resolved threat and its variants.

- **Strengthening Technical Controls**: Update firewalls, Intrusion Detection Systems (IDS), and software patches to address identified vulnerabilities. For instance, if malware exploited an unpatched server, future patching protocols should be revised to reduce lag time.

- **Scenario Planning**: Create simulated scenarios based on the threat to test and validate the updated response mechanisms.

Cultivating a Mindset of Resilience and Proactive Defense

In FinTech, where trust and data security are paramount, teams must adopt a forward-looking mindset that prioritizes resilience and anticipates emerging threats.

a. Promoting Resilience

- **Building Psychological Resilience**: Encourage an organizational culture that sees challenges as opportunities for growth. Recognize and reward team members who demonstrate initiative during incidents.

- **Iterative Learning**: Adopt a continuous feedback loop for training and threat readiness. For example, quarterly reviews of recent cybersecurity events can foster awareness and preparation.

- **Cross-Functional Collaboration**: Encourage cooperation between product, engineering, legal, and compliance teams. When everyone understands the impact of threats, responses become more cohesive and effective.

b. Establishing Proactive Defense

- **Threat Intelligence Integration**: Regularly incorporate insights from threat intelligence feeds to preemptively address emerging risks. For example, tracking phishing trends in the industry can help update email filters and employee training materials.

- **Penetration Testing and Ethical Hacking**: Simulate attacks on systems to identify vulnerabilities before real attackers can exploit them.

- **Automated Monitoring and Response**: Leverage AI-driven tools to monitor for anomalies and initiate automated responses to threats. For instance, anomaly detection can flag suspicious transaction patterns that deviate from norms.

Case Studies and Practical Exercises

Case Study 1: A FinTech Start-Up's Response to a Credential Stuffing Attack

A start-up providing digital wallets faced a credential stuffing attack, where hackers used stolen login credentials from previous breaches to access accounts.

- **Response**: The company detected unusual login attempts via monitoring tools and immediately implemented multi-factor authentication (MFA).

- **Post-Threat Analysis**: RCA revealed the lack of MFA as the primary vulnerability. The company also noted that user passwords weren't being evaluated for reuse across breached databases.

Improvements:

- MFA became mandatory for all users.

- Integrated Have I Been Pwned APIs to check new passwords against known breaches.

- Conducted user education campaigns on password hygiene.

- **Key Takeaway**: An agile response and comprehensive analysis turned a critical threat into an opportunity to reinforce trust and security.

Case Study 2: Insider Threat in a Payment Processing Firm

A payment processor discovered an employee accessing customer data without authorization.

- **Response**: The access was flagged by an anomaly detection system, and the individual's access was immediately revoked.

- **Post-Threat Analysis**: The review identified overly broad access permissions granted to the employee.

Improvements:

- Revised role-based access controls (RBAC) to follow the principle of least privilege.

- Enhanced employee background checks during hiring processes.

- Introduced real-time monitoring for sensitive data access.

- **Key Takeaway**: Addressing insider threats requires not just technical tools but robust governance and clear policies.

Practical Exercises

- **Simulated Incident Drill**: Conduct a tabletop exercise simulating a ransomware attack. Teams should walk through identification, containment, and recovery steps while evaluating the effectiveness of current playbooks.

- **Breach Scenario Training**: Provide employees with mock scenarios involving phishing emails, requiring them to identify and report threats.

- **Vulnerability Scanning Workshop**: Teach engineering teams to use tools like OWASP ZAP to identify application vulnerabilities proactively.

📖 Conclusion

Cybersecurity threats continue to evolve in complexity and scale, presenting significant challenges for organizations, particularly in the FinTech sector, where sensitive financial data and trust are paramount. This chapter emphasized the critical importance of proactive strategies, robust frameworks, and continuous improvement to address these challenges effectively.

By leveraging tools such as the MITRE ATT&CK framework, the NIST Cybersecurity Framework, and threat modeling techniques, organizations can adopt a structured approach to identifying vulnerabilities, mitigating risks, and responding to incidents. These methodologies not only strengthen defenses but also align security measures with operational goals and regulatory requirements, ensuring resilience in an ever-changing threat landscape.

Key takeaways from this chapter include the need for tailored security strategies, the integration of advanced technologies such as automation and AI-driven threat detection, and the cultivation of a culture of security and resilience across all levels of an organization. FinTech companies, in particular, can benefit from these insights by embedding security into every phase of their product lifecycle, from design to deployment, fostering trust among stakeholders and maintaining their competitive edge.

Ultimately, understanding and resolving cybersecurity threats is not a one-time effort but an ongoing process of learning, adaptation, and vigilance. By staying informed, investing in robust defenses, and fostering collaboration across teams and industries, organizations can not only protect themselves but also contribute to a safer digital ecosystem.

💡 What We Learnt:

- **Cybersecurity requires tailored approaches**: Each organization must design security strategies based on its unique vulnerabilities, adversarial behaviors, and operational needs.

- **The MITRE ATT&CK framework provides a detailed structure**: By categorizing tactics, techniques, sub-techniques, and procedures, organizations can map real-world adversarial behaviors to their systems and improve defenses.

- **FinTech organizations face unique risks**: Due to sensitive financial data and critical operations, they must focus on specific threats like API vulnerabilities, insider threats, and data breaches.

- **Threat modeling identifies risks early**: Incorporating threat modeling into product development ensures that vulnerabilities are addressed proactively, minimizing risks during the design phase.

- **The NIST Cybersecurity Framework offers flexibility**: Its five core functions—Identify, Protect, Detect, Respond, and Recover—can be tailored to align with organizational goals and regulatory requirements.

- **Incident response planning is crucial**: Detailed plans, regular simulations, and the use of automated tools improve the speed and effectiveness of responses to cybersecurity events.

- **Advanced tools enhance threat resolution**: Technologies like automated detection and response, incident management tools, and external threat intelligence feeds enable real-time threat mitigation.

- **Multifactor Authentication (MFA) and biometrics reduce identity theft risks**: These technologies provide additional layers of defense, making it harder for attackers to gain unauthorized access.

- **Proactive measures strengthen resilience**: Practices like penetration testing, ethical hacking, and continuous monitoring help organizations stay ahead of evolving threats.

- **Continuous improvement is key to cybersecurity**: Post-threat analysis, refinement of incident response plans, and the adoption of a proactive mindset ensure ongoing security and operational resilience.

PART 3
BUILDING SECURE FINTECH PRODUCTS

CHAPTER 6
SECURITY BY DESIGN

📖 Introduction

In today's digital economy, FinTech companies face constant cybersecurity challenges as they manage vast amounts of sensitive financial data and facilitate millions of daily transactions. With cyber threats evolving rapidly, securing financial platforms can no longer be a reactive process. This is where **Security by Design (SbD)**[130] comes into play—a proactive, systematic approach that embeds security into every stage of the product development lifecycle.

Rather than viewing security as a final checkpoint before a product launch, Security by Design ensures that cybersecurity principles are integrated from the very beginning. This approach mitigates risks, reduces costly breaches, and builds trust with customers, stakeholders, and regulators. By embedding robust security practices into the foundational layers of product development, FinTech companies can create systems that are resilient, scalable, and secure by default.

The importance of this approach becomes even clearer when considering the consequences of high-profile breaches in the FinTech sector. From compromised digital wallets and payment gateways to exposed personal data in investment platforms, the cost

---

[130] https://www.ikarussecurity.com/en/security-news-en/security-by-design-cybersecurity-throughout-the-product-life-cycle/

of inadequate security can be catastrophic. Companies that adopt a Security by Design mindset can prevent such incidents while ensuring compliance with global regulatory standards such as PCI DSS, GDPR, and PSD2.

This chapter explores how Security by Design strengthens FinTech[131] products by embedding security in the design process, integrating critical security features, and fostering a culture of continuous security awareness within product teams. It also outlines essential security measures such as role-based access control, data encryption, and secure API management. By the end of this chapter, you will understand how to transform security from a reactive task into a core business enabler, setting the stage for safer financial innovations.

📖 Embedding Security into the Product Development Lifecycle

Building secure FinTech products requires a systematic, proactive approach where security is integrated at every stage of the product development lifecycle. This ensures that potential vulnerabilities are addressed before they become critical issues. This approach includes phases such as requirement analysis, design, development, testing, deployment, and maintenance. Here's a detailed breakdown of how to embed security into each phase, complemented by relevant case studies:

---

[131]https://www.globalbankingandfinance.com/security-by-design-in-the-fintech-sector/

Requirement Analysis: Defining Security from the Beginning

Requirement analysis is the foundation of secure product development. This phase involves gathering all necessary information related to security policies, regulatory compliance, and business goals to define the product's security requirements.

Key Security Practices:
Regulatory Compliance:

- Identify applicable regulations such as GDPR (General Data Protection Regulation), PCI DSS (Payment Card Industry Data Security Standard), and PSD2 (Payment Services Directive 2).

- Conduct legal reviews to ensure compliance.

Business and Security Alignment:

- Align product features with business goals without compromising security.

- Define product objectives, including performance, reliability, and secure access.

Risk Assessment and Threat Identification:

- Perform a comprehensive risk assessment.

- Identify security threats such as data breaches, account takeovers, and insider threats.

Stakeholder Involvement:

- Involve all stakeholders, including developers, legal teams, product managers, and security specialists.

Case Study:
A global FinTech platform suffered multiple fines due to non-compliance with GDPR[132] data protection laws. After embedding regulatory compliance assessments in the requirement analysis phase, they reduced fines by 90% and improved data management practices[133].

Secure Design: Building Security into Product Architecture

Security design focuses on embedding security principles into the product's architecture and system workflows. This proactive approach prevents many vulnerabilities from arising later in the development lifecycle.

Core Design Principles:
Principle of Least Privilege:

- Ensure every user and system has only the minimum level of access needed.

- Apply role-based access controls (RBAC).

Data Security:

- Use encryption protocols such as AES-256 for data at rest and TLS for data in transit.

- Implement tokenization and hashing for sensitive data storage.

Authentication and Authorization:

---

[132] https://www.deskera.com/blog/role-based-access-control-power-financial-security/
[133] https://link.springer.com/article/10.1007/s12525-023-00622-x

- Implement multi-factor authentication (MFA), single sign-on (SSO), and OAuth 2.0 standards.

Threat Modeling:

- Conduct threat modeling to identify potential vulnerabilities.

- Develop mitigation strategies and integrate them into the system design.

Secure API Design:

- Secure APIs by applying rate limiting, authentication tokens, and HTTPS.

Case Study:

An e-commerce FinTech platform experienced significant account takeovers due to weak authentication protocols. After incorporating biometric authentication and adaptive security into its architecture, fraud incidents dropped by 75%[134].

Development Best Practices: Writing Secure Code

Secure coding is essential for minimizing vulnerabilities. Adopting industry-recognized standards and automating security processes reduces human errors and improves code quality.

Key Development Best Practices:
Follow Secure Coding Standards:

- Adhere to OWASP Secure Coding Guidelines and Common Weakness Enumeration (CWE).

- Avoid common vulnerabilities such as SQL injection, cross-site scripting (XSS), and buffer overflows.

---

[134] https://www.fintechfutures.com/2024/06/account-takeover-fraud-the-definitive-guide/

Use Secure Frameworks and Libraries:

- Rely on trusted, up-to-date libraries.

- Regularly monitor dependencies for vulnerabilities.

Implement Code Reviews and Audits:

- Conduct peer code reviews with a security focus.

- Use automated static application security testing (SAST) tools.

Version Control and Configuration Management:

- Use version control systems like Git to track changes.

- Manage configurations securely, avoiding hardcoded credentials.

Environment Isolation:

- Separate development, staging, and production environments to reduce risks.

Case Study:
A digital lending startup faced several service disruptions due to untracked code changes. After implementing automated code reviews and a centralized version control system, they improved development efficiency and reduced service downtime by 60%[135].

Testing and Validation: Ensuring Robust Security

Regular security testing ensures that vulnerabilities are detected and resolved before product release. Testing should be conducted

---

[135] https://cpostrategy.media/?p=interface

throughout the PDLC, from development to post-launch monitoring.

Security Testing Methods:
Static Application Security Testing (SAST):

- Analyze source code for security vulnerabilities during development.

Dynamic Application Security Testing (DAST):

- Test running applications to identify runtime vulnerabilities.

Penetration Testing:

- Simulate real-world attacks to identify exploitable weaknesses.

- Conduct both internal and external penetration testing.

Vulnerability Assessments:

- Perform continuous vulnerability scanning across all systems.

Manual and Automated Testing:

- Use manual testing for complex vulnerabilities and automated tools for routine checks.

Case Study:
A FinTech payment gateway detected critical API vulnerabilities during a routine penetration test, which helped prevent a major data breach. Post-testing, they implemented additional API security layers, reducing cyber-attack attempts by 80%[136].

---

[136] https://devcom.com/tech-blog/api-security-best-practices-protect-your-data/

Deployment and Maintenance: Sustaining Security Post-Launch

Security doesn't end after deployment. Continuous monitoring, timely updates, and proper incident response protocols ensure long-term protection against evolving threats.

Best Deployment and Maintenance Practices:

Secure Deployment Pipelines:

- Use infrastructure-as-code tools like Terraform and Docker for automated deployments.

- Implement continuous integration/continuous delivery (CI/CD) pipelines.

Regular Patching and Updates:

- Apply patches and updates immediately after security releases.

- Use automated patch management systems to reduce downtime.

Incident Response and Recovery Plans:

- Create and maintain an incident response plan (IRP).

- Conduct regular incident response simulations and tabletop exercises.

Monitoring and Logging:

- Use security information and event management (SIEM) tools for real-time monitoring.

- Implement anomaly detection systems powered by artificial intelligence.

Data Backups and Disaster Recovery:

- Schedule regular backups of critical data.

- Ensure off-site data replication for disaster recovery.

Case Study:

A banking application detected unauthorized access attempts through real-time log monitoring. By enabling automated responses and tightening its SIEM rules, the company mitigated the breach with zero customer impact[137].

Security as a Continuous Journey

Embedding security into the entire product development lifecycle ensures that FinTech products remain resilient against evolving cyber threats. By integrating security practices early and maintaining them continuously, businesses can build secure, scalable, and compliant financial solutions. This approach minimizes the risk of data breaches, regulatory penalties, and reputation damage, ultimately enhancing customer trust and long-term success.

Security Measures in Design and Architecture

In the rapidly evolving landscape of financial technology (FinTech), ensuring the security of digital platforms is paramount. With sensitive financial and personal data at stake, FinTech companies must adopt a robust, multi-layered security approach. This section outlines key security measures integral to FinTech system design and architecture, enhanced with real-world case studies to illustrate their practical application.

---

[137] https://www.upguard.com/blog/cyber-threat-detection-and-response

1. Role-Based Access Control (RBAC)

Role-Based Access Control (RBAC)[138] is a critical security mechanism that assigns permissions based on job roles. It ensures employees only access data and system components necessary for their tasks, reducing the risk of insider threats and accidental data exposure.

Case Study:

A global payment processing company implemented RBAC to manage user access across its distributed workforce. After adoption, the company reported a 35% decrease in security incidents stemming from unauthorized access. Their IT team also found it easier to manage and audit user permissions during compliance checks[139].

2. Data Encryption

Encryption secures sensitive data by converting it into unreadable formats, ensuring that even if data is intercepted, it remains protected. FinTech firms should encrypt data both at rest and in transit using industry-standard algorithms such as AES-256 and TLS.

Case Study:

A digital wallet provider faced a data breach in 2018 but minimized damage due to its robust encryption strategy. Despite attackers gaining access to some data, encryption rendered it useless, saving the company millions in potential losses and safeguarding customer trust[140].

---

[138] https://www.upguard.com/blog/rbac
[139] https://www.nist.gov/document/report02-1pdf
[140]
https://mrcet.com/pdf/Lab%20Manuals/IT/CYBER%20SECURITY%20(R18A0521).pdf

3. Multi-Factor Authentication (MFA)

Multi-Factor Authentication (MFA) adds multiple verification layers to user logins, significantly reducing the likelihood of unauthorized access. Common MFA methods include SMS codes, authenticator apps, and biometric verification like facial recognition or fingerprints.

Case Study:

A leading online banking platform experienced frequent account hijacking attempts. After introducing MFA, account takeovers dropped by over 80%. Customers reported enhanced trust in the platform, leading to increased engagement and service adoption[141].

4. Tokenization

Tokenization replaces sensitive data with unique identifiers or "tokens" during transactions, ensuring that critical information such as credit card numbers is never stored in its original form. This reduces data exposure and meets compliance requirements like PCI DSS.

Case Study:

A global e-commerce site adopted tokenization to secure customer payment data. After deployment, the company saw a 60% reduction in fraudulent payment attempts, while transaction processing speed improved due to streamlined token-based verification[142].

5. Secure APIs

APIs are essential for integrating third-party services in FinTech platforms. Using secure API gateways, OAuth standards, and

---

[141] https://darktrace.com/blog/two-factor-authentication-2fa-compromised-microsoft-account-takeover

[142] https://www.cybersource.com/content/dam/documents/campaign/fraud-report/global-fraud-report-2024.pdf

regular API auditing can prevent data leaks and unauthorized data access.

Case Study:

An investment platform suffered an API-based attack, resulting in unauthorized fund transfers. After integrating a secure API gateway and adopting OAuth 2.0, the platform enhanced its API security posture, preventing similar breaches and reinforcing customer confidence[143].

6. Zero Trust Architecture (ZTA)

Zero Trust Architecture (ZTA)[144] operates on the principle of "never trust, always verify." Every user, system, and device is considered potentially compromised. Continuous monitoring, real-time threat detection, and access control policies ensure maximum security.

Case Study:

A financial advisory firm adopted a Zero Trust model after experiencing repeated phishing attacks. By segmenting its network, enabling real-time monitoring, and applying strict user authentication policies, the firm reduced cyber threats by 75% within a year[145].

Conclusion

FinTech companies must implement these security measures to ensure comprehensive protection against modern cyber threats. Combining preventive and reactive strategies through RBAC, encryption, MFA, tokenization, secure APIs, and Zero Trust Architecture enables platforms to safeguard user data, maintain

---

[143]
https://www.researchgate.net/publication/386148601_Comprehensive_Framework_for_S ecuring_Financial_Transactions_through_API_Integration_in_Banking_Systems
[144]https://link.springer.com/chapter/10.1007/978-3-031-68005-2_7
[145] https://cdn-dynmedia-1.microsoft.com/is/content/microsoftcorp/microsoft/final/en-us/microsoft-brand/documents/Microsoft-Zero-Trust-TEI-Study.pdf?culture=en-us&country=us

regulatory compliance, and build long-term customer trust. Real-world examples demonstrate that investing in security is not only a technical necessity but also a business imperative in the competitive FinTech landscape.

Developing a Secure Culture Within Product Teams

Security is not just about tools—it's about people. A strong security culture within product teams ensures that security considerations are embedded in every stage of the product lifecycle. This involves educating teams, encouraging secure practices, fostering accountability, and promoting collaboration across functions. A secure culture helps mitigate risks, reduce vulnerabilities, and respond effectively to potential threats.

1. Security Training: Empowering Teams Through Knowledge

Regular cybersecurity awareness training is vital to equip product teams with the knowledge needed to identify, prevent, and respond to security threats. This includes educating teams on common attack vectors such as phishing, malware, and social engineering.

Case Study:
A global e-commerce platform experienced a surge in phishing attempts targeting its customer service team. After conducting quarterly cybersecurity training focused on identifying phishing emails, the company reported a 60% decrease in successful phishing incidents. Employees learned to recognize red flags, report suspicious activity, and follow secure protocols[146].

2. Security Champions: Driving Security from Within

Appointing security champions within product teams ensures continuous promotion of secure development practices. These

---

[146] https://aag-it.com/the-latest-phishing-statistics/

champions act as liaisons between product teams and the security department, advocating for security best practices during development cycles.

Case Study:
A fintech startup implemented a "Security Champion Program," assigning at least one trained security advocate per team. These champions ensured that secure coding practices were followed and that new features underwent security reviews. As a result, the company saw a 40% reduction in post-deployment vulnerabilities within a year[147].

3. Cross-Functional Collaboration: Security as a Shared Responsibility

Integrating security experts into product design discussions and development processes ensures that security is considered from the start. This prevents costly rework and reduces the likelihood of breaches due to overlooked vulnerabilities.

Case Study:
An IoT device manufacturer partnered with security consultants during the early stages of product design. By conducting regular security assessments and involving developers, testers, and UX designers, they reduced product recalls by 30% and maintained compliance with global security standards[148].

4. Incident Response Planning: Staying Prepared for the Worst

An incident response plan outlines clear steps for detecting, managing, and mitigating security breaches. Regular drills and

---

[147] https://www.temenos.com/wp-content/uploads/2024/04/Temenos_Sustainability-Report-2023.pdf

[148]

https://www.nist.gov/system/files/documents/2024/10/21/The%20IoT%20of%20Things%20Oct%202024%20508%20FINAL_1.pdf simulations ensure that teams know their roles during an actual incident, reducing response time and minimizing damage.

Case Study:
A healthcare software company developed a comprehensive incident response plan after experiencing a ransomware attack. They conducted monthly drills simulating various attack scenarios. When a real breach occurred, the team detected and contained the threat within hours, preventing sensitive data from being compromised[149].

5. Security Metrics: Measuring Success Through Data

Tracking security metrics helps organizations evaluate their security posture, identify areas for improvement, and demonstrate accountability. Key metrics include patch management timelines, detected vulnerabilities, resolved incidents, and time to incident resolution.

Case Study:
A SaaS provider implemented a security dashboard displaying real-time metrics such as patching timelines and open vulnerabilities. By monitoring these metrics, they identified delays in software patching and streamlined the process. This reduced their average patching time from 14 days to 4 days, lowering overall security risks.

Building a secure culture within product teams is a continuous process requiring education, collaboration, and accountability. By implementing training programs, appointing security champions, fostering cross-functional collaboration, conducting incident response drills, and tracking security metrics, organizations can

---

149

https://www.researchgate.net/publication/376514138_Ransomware_Attacks_on_Healthc are_Systems_Case_Studies_and_Mitigation_Strategies

create resilient teams capable of developing secure products while adapting to an ever-evolving threat landscape[150].

Conclusion

Security by Design is not just an option—it is a critical necessity in the FinTech industry. As the sector continues to evolve with rapid technological advancements, the importance of embedding security from the ground up cannot be overstated. A proactive security-first approach ensures that FinTech companies can build resilient, compliant, and trustworthy products that meet the complex demands of today's digital economy.

Incorporating security into every stage of the product development lifecycle minimizes vulnerabilities, reduces the risk of data breaches, and ensures compliance with stringent financial regulations. This includes conducting thorough security assessments, adopting secure coding practices, and continuously monitoring for emerging threats. By integrating security as a core component rather than an afterthought, organizations can stay ahead of potential attackers and respond effectively to evolving cyber threats.

Moreover, creating a strong security culture within teams fosters an environment where every individual understands their role in protecting sensitive data. Regular security training, appointing dedicated security champions, promoting cross-functional collaboration, and conducting incident response drills help establish a culture of shared responsibility. When employees are empowered

---

[150] https://www.researchgate.net/publication/346555570_Software_Security_Patch_Manage ment_--
_A_Systematic_Literature_Review_of_Challenges_Approaches_Tools_and_Practices

with knowledge and supported by a well-defined security framework, they become an organization's first line of defense.

Building secure products also strengthens customer trust, a crucial asset in the FinTech space. Customers expect their financial data to be handled securely, and a single security incident can severely damage a company's reputation. By adopting industry best practices such as encryption, multi-factor authentication, and secure API management, FinTech companies can demonstrate their commitment to data security and privacy.

Additionally, regulatory compliance is a non-negotiable aspect of operating in the FinTech industry. Meeting global and regional compliance standards such as PCI-DSS, GDPR, and ISO 27001 requires a well-defined security strategy. A robust security framework not only ensures regulatory compliance but also reduces legal liabilities and financial penalties associated with data breaches.

The cost of neglecting security can be catastrophic, leading to financial losses, operational disruptions, and loss of customer confidence. On the other hand, investing in a comprehensive security strategy enables long-term business sustainability, operational resilience, and market competitiveness. In a world where cyber threats are becoming more sophisticated, only companies that prioritize Security by Design will be equipped to navigate the digital economy securely.

In conclusion, FinTech companies must view Security by Design as a strategic investment rather than a technical expense. By embedding security into the core of their products and fostering a culture of continuous improvement, they can build a safer, more secure, and resilient financial ecosystem. This proactive approach ensures not only business success but also contributes to a more trustworthy and sustainable digital economy for all stakeholders.

What We Learnt:

- **Proactive Security Integration:** Security should be embedded throughout the product development lifecycle to prevent vulnerabilities.

- **Regulatory Compliance:** Adhering to standards like PCI DSS, GDPR, and PSD2 helps avoid legal issues and fines.

- **Risk Assessment and Threat Identification:** Identifying potential risks early minimizes the chances of data breaches.

- **Core Security Principles:** Implementing least privilege, data encryption, multi-factor authentication (MFA), and secure APIs strengthens system security.

- **Secure Product Development:** Following secure coding practices, conducting code reviews, and using automated testing ensures robust application security.

- **Continuous Security Monitoring:** Regular patching, real-time monitoring, and incident response plans maintain system resilience post-launch.

- **Security Culture in Teams:** Promoting security awareness, appointing security champions, and fostering cross-functional collaboration enhances organizational security readiness.

- **Business Trust and Customer Confidence:** A strong security posture builds customer trust, reduces reputational damage, and ensures long-term business sustainability.

CHAPTER 7
AUTHENTICATION AND USER SECURITY

Ensuring the security of users' data and digital identities is a critical component of modern application development. Authentication and user security mechanisms prevent unauthorized access and safeguard sensitive information. This chapter explores best practices for implementing strong authentication methods, managing user permissions, mitigating social engineering attacks, and balancing security with a seamless user experience (UX).

📖 Implementing Strong Authentication

Authentication is the process of verifying a user's identity before granting access to a system. To ensure maximum security, organizations must adopt multi-layered authentication mechanisms.

Two-Factor Authentication (2FA)

Two-factor authentication (2FA)[151] is a security process that strengthens user authentication by requiring two separate forms of verification before granting access to a system, application, or online service. This additional layer of security helps protect sensitive data and accounts from unauthorized access, even if one

---

[151] https://www.consumerreports.org/electronics/digital-security/best-way-to-use-two-factor-authentication-a1070812070/

authentication factor is compromised. The two factors used in 2FA typically fall into the following categories:

Something They Know:

This includes knowledge-based credentials such as passwords, PINs, or answers to security questions. For example, when logging into an email account, users must first enter their password. This is the most common authentication factor, but it is also the most vulnerable due to phishing, password guessing, and data breaches.

Something They Have:

This involves possession-based items that users must physically have with them, such as:

- **Smartphones:** Used for receiving SMS codes, push notifications, or generating time-based one-time passwords (TOTP) through apps like Google Authenticator, Microsoft Authenticator, or Authy.

- **Hardware Tokens:** Dedicated security devices like YubiKeys or RSA SecureID tokens that generate unique codes or connect via USB for authentication.

- **Smart Cards:** Common in corporate environments, smart cards store authentication credentials securely.

Something They Are:

This factor relies on biometrics — unique physical or behavioral traits of the user. Examples include:

- **Fingerprints:** Widely used in smartphones, laptops, and secure entry systems.

- **Facial Recognition:** Found in many modern devices like iPhones with Face ID and Windows Hello-enabled computers.

- **Iris Scans or Voice Recognition:** Employed in high-security environments requiring advanced biometric verification.

Popular 2FA Methods:

Several 2FA methods are widely used across various platforms and services:

- **Time-Based One-Time Passwords (TOTP):** Generated by authenticator apps such as Google Authenticator or Authy, TOTP codes refresh every 30 seconds, making them secure and difficult to guess.

- **SMS Codes:** While still commonly used, SMS-based 2FA is considered less secure due to the risk of SIM-swapping attacks and message interception.

- **Push Notifications:** Trusted apps like Duo Security and Microsoft Authenticator send real-time push notifications, allowing users to approve or deny login attempts with a single tap.

Benefits of 2FA:

Implementing 2FA significantly reduces the likelihood of account compromise, even if passwords are leaked or stolen. It creates an additional barrier that cybercriminals must bypass, enhancing security for both individuals and organizations. Many online services, including banking apps, social media platforms, and email providers, now recommend or mandate 2FA for added protection.

🔍 **Real-World Example:** Consider an employee accessing a corporate network remotely. After entering their work email and password, they receive a push notification on their smartphone from an authenticator app. They must tap "Approve" to complete the login process. Even if a malicious actor knows the employee's password, access would still be denied without control of the registered smartphone.

By leveraging 2FA, organizations can minimize security breaches, ensure compliance with industry standards, and provide users with greater peace of mind.

Managing User Permissions and Mitigating Social Engineering Attacks

Proper user access management is a critical component of cybersecurity, designed to minimize security risks by ensuring that individuals only have access to the data and systems necessary for their specific job functions. Mismanaged permissions can lead to data breaches, insider threats, and increased vulnerability to external attacks. Implementing robust access management protocols helps maintain the integrity, confidentiality, and availability of an organization's resources.

Role-Based Access Control (RBAC)

Role-Based Access Control (RBAC) is a widely adopted access management strategy that assigns permissions based on predefined roles within an organization. These roles are determined by job responsibilities, ensuring that employees have the necessary permissions to perform their tasks without gaining unnecessary access to sensitive data.

🔍 **Example:** In a hospital setting, doctors may have access to patient records and medical histories, while administrative staff may only access appointment schedules and billing information. This segmentation prevents unauthorized access to sensitive medical records, ensuring data privacy and compliance with regulations like HIPAA.

Best Practices for Implementing RBAC:

- **Define Clear Roles:** Establish roles that align with job functions, avoiding overly broad categories.

- **Regularly Review Permissions:** Conduct periodic audits to ensure permissions reflect current job responsibilities.

- **Apply the Principle of Least Privilege:** Limit access to only what is necessary for each role to reduce the attack surface.

Principle of Least Privilege (PoLP)

The Principle of Least Privilege (PoLP) ensures that users are granted the minimum permissions required to perform their tasks. This limits the impact of potential breaches, compromised accounts, or insider threats.

🔍 **Example:** An intern in an IT department might only be allowed to view support tickets but not modify system configurations. If the intern's account is compromised, the attacker's access is restricted, minimizing potential damage.

PoLP Implementation Strategies:

- **Role Segmentation:** Break down roles into granular permissions.

- **Just-in-Time Access:** Provide temporary elevated permissions only when required for specific tasks.

- **Access Reviews and Audits:** Conduct frequent reviews to remove unnecessary permissions and detect anomalies.

Mitigating Social Engineering Attacks

Social engineering attacks exploit human psychology to trick individuals into revealing confidential information, bypassing technological security measures. Common tactics include phishing emails, pretexting, baiting, and tailgating.

Types of Social Engineering Attacks:

- **Phishing:** Sending fraudulent emails that appear to come from trusted sources to steal sensitive data.

- **Pretexting:** Creating a false scenario to trick someone into divulging information.

- **Baiting:** Offering something enticing, such as free software downloads, to lure victims into installing malware.

- **Tailgating:** Following authorized personnel into restricted areas without proper authentication.

Best Practices for Mitigating Social Engineering:

- **Regular Security Training:** Conduct frequent security awareness sessions, covering common social engineering tactics and how to recognize them.

- **Verification Processes:** Implement multi-factor authentication (MFA) and verification steps before granting access to sensitive systems.

- **Phishing Simulations:** Conduct simulated phishing attacks to test employee readiness and improve incident response.

- **Incident Reporting Mechanisms:** Establish clear reporting procedures for suspected attacks to ensure quick response and containment.

- **Automated Monitoring and Anomaly Detection:** Deploy tools that monitor user activities for unusual behavior, such as login attempts from unfamiliar locations or unexpected data access patterns.

🔍 **Example:** An organization may send a simulated phishing email to employees, offering a fake gift card. Employees who click the link are redirected to a training module that explains how to spot phishing attempts, reinforcing learning through practical experience.

By combining access management strategies like RBAC and PoLP with comprehensive training and proactive monitoring, organizations can effectively reduce security risks and strengthen their defenses against social engineering attacks.

Balancing Security with User Experience (UX)

In today's digital age, security is essential to protect user data and maintain trust. However, enhancing security should not compromise user experience (UX)[152]. A balance must be struck where users feel secure without being overwhelmed by complicated security protocols. Poorly designed security measures can frustrate users, leading to reduced adoption rates, abandoned services, or the development of risky workarounds that undermine overall security. Therefore, modern systems must integrate robust security features while maintaining an intuitive and user-friendly interface.

---

[152] https://blog.1password.com/authentication-methods/

Designing User-Centered Security Systems

Designing security systems that prioritize user experience involves creating intuitive, seamless, and effective security processes. This approach reduces user frustration and enhances system adoption while maintaining a high level of security. Here are some key strategies:

Simplify Authentication Flows

Simplifying how users authenticate can reduce friction and enhance usability. Examples include:

- **Single Sign-On (SSO):** This method allows users to log in once and gain access to multiple services. For instance, logging into a Google account also grants access to Gmail, Google Drive, and other related services.

- **Password Managers:** These tools securely store complex passwords, allowing users to log in automatically without remembering multiple credentials. Popular password managers like LastPass and Bitwarden provide encrypted storage and autofill capabilities.

Minimize Friction with Modern Authentication Methods

Traditional passwords can be cumbersome and prone to breaches. Modern alternatives improve both security and UX:

- **Biometric Authentication:** Methods like fingerprint scanning, facial recognition, and iris detection enable secure and seamless logins. For example, Apple's Face ID and Touch ID systems simplify device and app access while maintaining high-security standards.

- **Two-Factor Authentication (2FA):** Adding an extra verification step, such as a code sent to a user's phone, enhances security while keeping processes straightforward. Google Authenticator and Authy are popular examples.

Offer Clear and Secure Account Recovery Options

Account recovery is crucial for users who forget their login credentials. A well-designed recovery process balances simplicity and security by:

- Allowing password resets through secure email links.

- Implementing multi-step verification, such as answering security questions or providing identification documents.

- Offering backup codes or recovery keys, often used by services like Microsoft and GitHub.

Continuous Authentication

Continuous authentication involves verifying a user's identity throughout their session, ensuring security without repeated logins. Unlike traditional methods requiring authentication only at login, continuous authentication keeps monitoring user behavior, adding an extra layer of security.

How Continuous Authentication Works:

- **Behavioral Biometrics:** Systems analyze users' behavioral patterns, such as typing speed, keystroke dynamics, or mouse movement patterns. For instance, banks often use these features to detect potential fraud.

- **Geolocation and Device Recognition:** Monitoring a user's physical location and the device they use can identify suspicious activities, such as login attempts from unfamiliar locations or unregistered devices.

- **Session Monitoring:** Continuous checks ensure that the person using a session remains the authorized user. If anomalies are detected, the system can trigger additional verification or end the session.

🔍 Examples of Continuous Authentication:

- **E-commerce Websites:** To prevent fraud, some platforms monitor user behavior, such as browsing patterns and purchase histories, flagging unusual activities.

- **Corporate Security Systems:** Companies use tools like Microsoft Azure Active Directory to enforce conditional access based on real-time user behavior, minimizing breaches while keeping the experience smooth for employees.

By combining these strategies, organizations can build systems that maintain high security standards while ensuring a smooth, frustration-free user experience. Striking this balance ultimately leads to increased trust, user retention, and long-term business success.

Conclusion

Implementing robust authentication and user security practices is essential for protecting digital assets and safeguarding user privacy in today's interconnected digital landscape. As cyber threats become increasingly sophisticated, organizations must adopt a multi-layered security approach to minimize vulnerabilities and prevent unauthorized access.

Strong authentication methods, such as multi-factor authentication (MFA), biometric verification, and advanced encryption protocols, play a critical role in ensuring secure access to sensitive information.

Additionally, managing user permissions effectively by adhering to the principle of least privilege can help reduce the attack surface by limiting users' access to only what is necessary for their roles.

Organizations should also take proactive measures to mitigate social engineering threats, such as phishing and impersonation attacks, by fostering a culture of cybersecurity awareness through regular training and simulations. Employees, customers, and partners must be educated on recognizing potential risks and following best security practices to prevent data breaches.

Moreover, while implementing these security measures, maintaining a balance between security and user experience is essential. Overly complex authentication processes can frustrate users and hinder productivity, while too lenient measures may expose the system to cyberattacks. A well-designed security framework should provide seamless access for legitimate users while keeping malicious actors at bay.

By consistently reviewing and updating security policies, monitoring system activity for unusual behavior, and staying informed about emerging cybersecurity trends, organizations can create a resilient digital environment. These efforts not only protect valuable data but also build trust and credibility among users, fostering long-term success in an increasingly digital world.

What We Learnt:

- **Protecting user data and digital identities is crucial** in modern application development to prevent unauthorized access and safeguard sensitive information.

- **Authentication involves verifying a user's identity** before granting access to a system, ensuring only authorized users can proceed.

- **Two-Factor Authentication (2FA) strengthens security** by requiring two separate verification methods, such as passwords and SMS codes or biometrics.

- **Common 2FA methods include TOTP, SMS codes, and push notifications**, enhancing account security even if one authentication factor is compromised.

- **Role-Based Access Control (RBAC) assigns permissions** based on predefined roles, limiting users' access to only what is necessary for their tasks.

- **The Principle of Least Privilege (PoLP) ensures minimum access** by granting users only the permissions required for their roles, reducing potential security breaches.

- **Social engineering attacks exploit human behavior**, using tactics like phishing, pretexting, and baiting to trick users into sharing sensitive information.

- **Security training and phishing simulations help mitigate attacks** by increasing user awareness and readiness to recognize and respond to threats.

- **Balancing security with user experience (UX) is essential** to ensure users feel secure without being overwhelmed by complex security protocols.

- **Modern authentication methods like biometrics and password managers** improve both security and user convenience by simplifying login processes.

- **Continuous authentication verifies users throughout a session,** monitoring behaviors like typing patterns and device recognition to detect suspicious activity.

- **Regular security policy updates and system monitoring** help detect and respond to emerging threats, building a secure and trustworthy digital environment.

CHAPTER 8
VULNERABILITY TESTING AND CONTINUOUS MONITORING

📖 Introduction

In the ever-evolving landscape of cybersecurity, vulnerability testing and continuous monitoring form the backbone of a robust security strategy. Organizations must identify potential weaknesses in their systems and respond to threats in real-time to minimize damage. This chapter explores key techniques for vulnerability assessment, the process of continuous monitoring, and the growing importance of bug bounty programs.

📖 Techniques for Vulnerability Assessment

Vulnerability assessment[153] is a systematic process that identifies, quantifies, and ranks vulnerabilities in an IT system. It includes various techniques such as penetration testing, security audits, and automated scanning. Each method has distinct features, advantages, and challenges.

---

[153] https://www.eccouncil.org/cybersecurity-exchange/penetration-testing/penetration-testing-best-practices-enterprise/

1. Penetration Testing

Penetration testing, commonly referred to as pen testing, is a proactive cybersecurity assessment method designed to simulate real-world cyberattacks on an organization's digital infrastructure. This process involves evaluating the security of systems, networks, applications, and other IT assets to identify vulnerabilities that could be exploited by malicious attackers.

Certified ethical hackers, also known as penetration testers, are professionals trained to think like cybercriminals while adhering to legal and ethical standards. They use advanced tools and techniques to uncover weaknesses in a company's defenses. By exploiting these vulnerabilities in a controlled and safe environment, they can determine how well an organization's security protocols stand up against potential threats.

The primary objective of penetration testing is to strengthen an organization's security posture by revealing hidden risks before they can be exploited by attackers. This includes identifying system misconfigurations, outdated software, weak passwords, and coding flaws. After completing the assessment, penetration testers provide a comprehensive report detailing discovered vulnerabilities, the methods used to exploit them, and actionable recommendations for improving overall cybersecurity defenses.

Penetration testing can be performed internally or externally, depending on the scope of the test. Internal testing simulates an attack from within the organization, mimicking a scenario where a disgruntled employee or someone with insider access causes harm. External testing focuses on threats originating outside the organization, such as hackers attempting to breach the company's public-facing systems.

Overall, penetration testing plays a crucial role in a robust cybersecurity strategy by ensuring that an organization's defenses are up-to-date, resilient, and capable of mitigating real-world cyber threats.

Types of Penetration Testing:

- **Black-box Testing:** Testers have no prior knowledge of the system.

- **White-box Testing:** Testers have full knowledge of the system's architecture.

- **Gray-box Testing:** Testers have limited knowledge, simulating insider threats.

🔍 **Example:** Consider a financial institution conducting a black-box pen test on its online banking platform. Testers discover a vulnerability in the login process that allows brute-force attacks. Mitigating this issue involves implementing rate-limiting and multi-factor authentication.

Best Practices:

- **Conduct Tests Regularly:** Perform routine security assessments, including penetration testing, vulnerability scanning, and system audits. This helps identify potential risks early and ensures that systems remain resilient against emerging threats. Schedule tests periodically and after major system updates or changes.

- **Engage Third-Party Professionals:** Collaborate with external cybersecurity experts to gain an unbiased assessment of your organization's security posture. Third-party professionals bring specialized skills, advanced tools, and industry best practices that can uncover vulnerabilities internal teams might miss.

- **Document Findings and Remediate Vulnerabilities:** Maintain detailed records of all security assessments, including identified vulnerabilities, the severity of issues, and recommended solutions. Use these findings to implement corrective actions promptly. Ensure a structured approach to remediation by assigning responsibilities, setting deadlines, and verifying that vulnerabilities are fully addressed through follow-up testing.

2. Regular Security Audits

Security audits are systematic evaluations designed to assess the effectiveness and robustness of an organization's security policies, processes, and infrastructure. These audits involve a thorough review of the organization's current security practices to identify potential vulnerabilities, gaps, and areas for improvement. They help ensure that the organization's operations align with established industry standards, regulatory frameworks, and legal requirements, minimizing the risk of data breaches and cyber threats.

During a security audit, auditors examine various components of the organization's security architecture, including network security, access controls, data protection protocols, and incident response procedures. They also evaluate the implementation of security policies, employee awareness training, and the effectiveness of cybersecurity technologies in place.

The audit process typically includes reviewing system configurations, analyzing access logs, conducting penetration testing, and verifying compliance with relevant security frameworks such as ISO 27001, PCI-DSS, or HIPAA, depending on the industry.

Auditors document their findings in detailed reports, highlighting any weaknesses discovered and offering recommendations for remediation.

By conducting regular security audits, organizations can proactively address vulnerabilities, enhance their overall security posture, and maintain trust with customers, partners, and regulatory bodies. These audits play a critical role in safeguarding sensitive information, ensuring business continuity, and reducing the likelihood of financial and reputational damage caused by security breaches.

Internal vs. External Audits:

- **Internal Audits:** Conducted by in-house teams.

- **External Audits:** Conducted by third-party experts for unbiased assessments.

🔍 **Example:** A healthcare provider conducts annual external audits to ensure compliance with the Health Insurance Portability and Accountability Act (HIPAA). The auditors identify outdated software in the patient records system, prompting an immediate upgrade.

Best Practices:
- **Establish a Regular Audit Schedule:** Develop a comprehensive audit calendar that includes periodic reviews of systems, processes, and policies. This should cover internal audits, external assessments, and compliance checks. A consistent audit schedule helps identify gaps, ensures regulatory compliance, and keeps the organization proactive rather than reactive to potential risks.

- **Involve Cross-Functional Teams:** Engage diverse teams from various departments, including IT, operations, legal, and compliance, to ensure a well-rounded approach to risk management. Cross-functional collaboration promotes a broader understanding of potential vulnerabilities and enables the implementation of more effective, organization-wide solutions.

- **Prioritize Remediation Based on Risk Severity:** Use a risk-based approach to remediation by assessing the potential impact and likelihood of identified vulnerabilities. Address critical and high-risk issues first to minimize exposure. Develop a systematic process that includes defining action plans, assigning responsibilities, and setting timelines to ensure timely and efficient mitigation.

3. Automated Vulnerability Scanning

Automated tools play a crucial role in enhancing IT system security by conducting comprehensive scans for known vulnerabilities. These tools work by systematically analyzing system configurations, software versions, and security settings across an organization's IT infrastructure. They compare the collected data against an extensive database of known threats, including vulnerabilities, misconfigurations, and outdated software components.

The database of known threats is regularly updated with information from various trusted security sources, including vulnerability repositories, security advisories, and vendor updates. When the tools detect a match between the system's current state and a documented vulnerability, they generate detailed reports highlighting the specific issues, their severity levels, and potential impact on the organization's security posture.

In addition to identifying vulnerabilities, automated scanning tools often provide remediation recommendations, helping IT teams prioritize and address security gaps efficiently. Some advanced tools may also support automatic patching or integration with security information and event management (SIEM) systems to streamline the incident response process.

By performing scheduled or continuous scans, these tools ensure that organizations remain proactive in managing security risks, minimizing the window of exposure to potential cyberattacks. This automated approach significantly reduces the manual workload on IT security teams, allowing them to focus on more complex and strategic security tasks.

Popular Tools:

- Nessus
- OpenVAS
- QualysGuard

🔍 **Example:** A retail company deploys an automated vulnerability scanner to detect outdated plugins on its e-commerce website. The scanner alerts the IT team, enabling them to update the plugins before cybercriminals can exploit them.

Best Practices:

- **Integrate Automated Scans into Development Pipelines:** Incorporate automated security scanning tools into the software development lifecycle (SDLC) to detect vulnerabilities early. This includes static application security testing (SAST) for code analysis, dynamic application security testing (DAST) for runtime evaluation, and software composition analysis (SCA) for third-party dependency checks. Automated scans should be triggered at key stages, such as during code commits, builds, and releases, ensuring continuous security validation without slowing down development.

- **Update Scanner Tools Regularly:** Keep security scanning tools up to date with the latest definitions, plugins, and threat intelligence feeds. Cybersecurity threats evolve rapidly, and outdated tools may miss emerging vulnerabilities. Regular updates ensure that scanners remain effective against newly discovered exploits and align with current industry standards and compliance requirements.

- **Review Scan Results Thoroughly:** Conduct detailed reviews of scan reports to identify, categorize, and prioritize vulnerabilities based on risk severity and business impact. Involve security specialists and development teams in the review process to ensure accurate assessment and appropriate remediation planning. Address false positives through manual verification and refine scanning rules as needed to improve accuracy over time. Implement follow-up scans after remediation to confirm that vulnerabilities have been successfully resolved.

📖 Continuous Monitoring and Rapid Response

Continuous monitoring[154] ensures that IT systems remain secure by detecting, reporting, and responding to threats in real-time. This proactive approach helps minimize potential damage from security incidents.

Key Elements of Continuous Monitoring

- **Data Collection:** Gather data from network devices, applications, and endpoints.

- **Analysis:** Use advanced analytics to detect anomalies and suspicious activities.

- **Alerting:** Notify security teams about potential threats.

- **Incident Response:** Implement pre-defined procedures to mitigate incidents quickly.

🔍 **Example:** A multinational corporation employs a Security Information and Event Management (SIEM) system for continuous monitoring. When unusual login attempts are detected from a foreign IP address, the system triggers an alert, prompting the security team to block the IP and investigate.

Best Practices:

- **Use SIEM Solutions for Centralized Monitoring:** Deploy a Security Information and Event Management (SIEM)

---

[154] https://bluegoatcyber.com/blog/red-team-vs-blue-team-cybersecurity-exercises-explained/

system to centralize the collection, monitoring, and analysis of security events from various sources, such as firewalls, servers, and endpoint devices. SIEM solutions provide real-time visibility into network activity, enabling the detection of suspicious behavior and potential breaches. Ensure that the SIEM platform is properly configured with relevant log sources, correlation rules, and threat intelligence feeds for optimal performance and accurate threat detection.

- **Establish a 24/7 Security Operations Center (SOC):** Build or partner with a dedicated Security Operations Center (SOC) that operates around the clock to monitor, detect, and respond to security incidents. A 24/7 SOC ensures continuous surveillance of the organization's digital infrastructure, reducing response times and minimizing the impact of potential threats. Equip the SOC team with advanced tools, up-to-date threat intelligence, and well-defined escalation procedures for handling incidents effectively.

- **Conduct Regular Incident Response Drills:** Perform scheduled incident response simulations and tabletop exercises to test the organization's readiness for various cyber threats, including ransomware attacks, data breaches, and denial-of-service (DoS) attacks. These drills help assess the effectiveness of the incident response plan, identify gaps, and refine processes accordingly. Involve all relevant stakeholders, including IT, legal, communications, and management teams, to ensure a coordinated and efficient response during real incidents.

Rapid Response Strategies

Containment:

- Identify and isolate affected systems quickly to prevent the malicious activity from spreading to other parts of the network.

- Disconnect compromised devices from the network, disable user accounts involved in the breach, and block malicious IP addresses or domains.

- Implement network segmentation and apply firewalls or access controls to limit the attacker's movement.

Eradication:

- Conduct a thorough investigation to identify all malicious components, including malware, backdoors, and unauthorized access points.

- Remove or neutralize all detected threats using antivirus, anti-malware tools, and manual intervention when necessary.

- Patch vulnerabilities, update system configurations, and change compromised credentials to prevent reinfection.

Recovery:

- Restore affected systems from clean backups, ensuring they are free from malicious components.

- Test restored systems thoroughly to confirm that they function as expected and that no residual threats remain.

- Monitor systems closely for signs of recurrence and maintain heightened security vigilance during the recovery phase.

Post-Incident Review:

- Conduct a comprehensive analysis of the incident to determine its root cause, entry points, and the extent of the damage.

- Review the effectiveness of the response process, including how well detection, containment, eradication, and recovery steps were executed.

- Update and improve the incident response plan based on lessons learned, incorporating new security measures, employee training, and process adjustments to reduce the risk of future incidents.

📖 Bug Bounty Programs and Their Benefits

Bug bounty programs[155] incentivize security researchers and ethical hackers to identify and report vulnerabilities in exchange for rewards. These programs enhance security by leveraging a global pool of skilled professionals.

How Bug Bounty Programs Work

- **Program Launch:** Organizations define the program scope and set reward amounts.

- **Vulnerability Reporting:** Researchers submit vulnerability reports.

- **Verification:** Security teams validate reported vulnerabilities.

---

[155] https://iterasec.com/blog/red-team-vs-blue-team-how-they-help-each-other/

- **Reward Distribution:** Valid submissions receive monetary rewards or recognition.

🔍 **Example:** A tech company launches a bug bounty program for its mobile app. A researcher discovers a critical authentication flaw and submits a detailed report. The company fixes the flaw and rewards the researcher with $10,000.

Benefits of Bug Bounty Programs

Enhanced Security: Bug bounty programs allow organizations to identify and address vulnerabilities in their systems before malicious actors can exploit them. By leveraging the expertise of ethical hackers from around the world, companies can uncover a wide range of security issues, including those that might be missed during traditional security assessments. This proactive approach helps strengthen the overall security posture of an organization by continuously testing and improving its defenses.

Cost-Effectiveness: Compared to hiring a dedicated in-house security team, bug bounty programs offer a more economical alternative. Organizations only pay for verified vulnerabilities that researchers discover, ensuring that resources are allocated efficiently. This "pay-for-results" model eliminates the need for maintaining full-time security testers, reducing operational costs while still providing high-quality security testing from diverse experts.

Community Engagement: Bug bounty programs foster collaboration with the global cybersecurity community. Ethical hackers gain recognition, rewards, and a platform to showcase their skills. This creates a mutually beneficial relationship where researchers receive incentives, and organizations benefit from their expertise. Engaging with the community also helps build trust and transparency, showing that the organization values security and is committed to protecting its users.

Access to a Diverse Talent Pool: Bug bounty programs attract cybersecurity researchers from various backgrounds and levels of expertise. This diverse group brings unique perspectives and problem-solving approaches, increasing the chances of uncovering complex and hard-to-find vulnerabilities that internal teams might overlook. The collective intelligence of a global network of researchers significantly enhances the effectiveness of security testing.

Improved Product Quality: Beyond security, bug bounty programs contribute to overall product improvement. By identifying bugs and vulnerabilities early in the development lifecycle, companies can refine their software, reduce system downtime, and enhance user experience. This results in more reliable products and services, boosting customer satisfaction and brand reputation.

Compliance and Risk Management: Many industry regulations require organizations to conduct regular security assessments. Bug bounty programs can help meet these compliance standards by providing continuous vulnerability testing. This not only reduces legal and financial risks but also demonstrates a commitment to maintaining the highest security standards, which can be valuable when working with partners and customers.

Brand Reputation and Trust: Running a bug bounty program signals that an organization is serious about cybersecurity and is willing to invest in safeguarding its digital assets. This transparency can enhance public trust and strengthen brand reputation. Customers and stakeholders are more likely to engage with a company that prioritizes security and takes proactive measures to protect their data.

Best Practices:

- Clearly define program rules and scope.

- Ensure timely responses to reported vulnerabilities.

- Offer competitive rewards to attract top talent.

📖 Advanced Concepts in Vulnerability Testing

Vulnerability testing is a critical aspect of cybersecurity, designed to identify and address potential weaknesses in systems, networks, and applications before they can be exploited by malicious actors. As cyber threats evolve, advanced concepts in vulnerability testing have emerged, focusing on proactive measures, real-time attack simulations, and collaborative defense strategies.

Red Team vs. Blue Team Exercises

Red Team vs. Blue Team exercises[156] are structured simulations used to evaluate and enhance an organization's cybersecurity posture. These exercises simulate real-world attack scenarios, enabling security teams to practice detection, response, and mitigation.

- **Red Team:** Composed of offensive security experts who emulate real-world attackers. Their role is to find vulnerabilities by conducting sophisticated attacks,

---

[156] https://www.sprocketsecurity.com/resources/red-team-vs-blue-team-roles-skills-tools-and-tips

including social engineering, penetration testing, and network exploitation. They think like hackers to uncover security gaps that may otherwise go unnoticed.

- **Blue Team:** Responsible for defending the organization's digital assets. They monitor networks, implement security protocols, and respond to simulated breaches. Their objective is to detect, analyze, and neutralize threats in real-time, ensuring minimal damage and disruption.

- **Purple Team:** A relatively newer concept, the Purple Team bridges the gap between Red and Blue teams by fostering collaboration. They facilitate knowledge sharing, enabling both teams to learn from each other's strategies, thereby enhancing the overall security posture.

These exercises improve threat detection capabilities, reduce response times, and ensure that defensive measures are continuously refined to counter emerging threats.

Zero-Day Exploit Testing

Zero-day exploits[157] pose some of the most significant threats to modern organizations. These vulnerabilities are unknown to vendors and remain unpatched, making them prime targets for attackers.

- **Zero-Day Vulnerabilities:** These are security flaws that attackers exploit before the software developers are aware of them. Zero-day attacks can cause substantial financial and reputational damage due to the lack of immediate fixes.

- **Testing Methods:** To combat zero-day threats, organizations employ advanced penetration testing, vulnerability assessments, and threat simulations. These

---

[157] https://iterasec.com/blog/red-team-vs-blue-team-how-they-help-each-other/

techniques include fuzz testing, reverse engineering, and threat modeling to uncover vulnerabilities proactively.

🔍 **Example:** Consider a major tech corporation that runs zero-day simulations on its cloud infrastructure. During one such exercise, its Red Team identifies a severe data breach vulnerability related to insecure API configurations. By conducting this test before the vulnerability is publicly known, the organization patches the flaw, preventing potential exploitation.

📖 Conclusion

Vulnerability testing and continuous monitoring are essential components of a modern cybersecurity strategy. Techniques such as penetration testing, security audits, and automated scanning help identify system weaknesses, while continuous monitoring ensures real-time threat detection. Additionally, bug bounty programs engage the broader cybersecurity community, creating a win-win scenario for organizations and security researchers. By implementing these practices, organizations can build resilient systems capable of withstanding the ever-growing range of cyber threats.

💡 What We Learnt:

- **Vulnerability Assessment:** Involves identifying, quantifying, and ranking vulnerabilities using techniques such as penetration testing, security audits, and automated scanning.

- **Penetration Testing:** Simulates real-world attacks to identify and exploit vulnerabilities, helping organizations strengthen their cybersecurity defenses.

- **Types of Penetration Testing:** Includes black-box (no prior knowledge), white-box (full system knowledge), and gray-box (limited knowledge) testing.

- **Regular Security Audits:** Evaluate the effectiveness of security policies and systems, ensuring compliance with industry standards and regulations.

- **Automated Vulnerability Scanning:** Uses tools to detect system vulnerabilities, misconfigurations, and outdated software, enabling quick remediation.

- **Continuous Monitoring:** Involves real-time data collection, analysis, alerting, and incident response to detect and mitigate threats proactively.

- **Rapid Response Strategies:** Focus on containment, eradication, recovery, and post-incident reviews to minimize damage from cyberattacks.

- **Bug Bounty Programs:** Encourage security researchers to report vulnerabilities in exchange for rewards, enhancing overall system security.

- **Advanced Testing Concepts:** Include Red Team vs. Blue Team exercises, zero-day exploit testing, and the introduction of Purple Teams for collaborative defense.

- **Best Practices:** Emphasize conducting regular tests, engaging third-party experts, maintaining updated tools, and documenting findings for continual improvement.

- **Overall Importance:** Implementing these techniques ensures a resilient cybersecurity posture, reducing the risk of financial and reputational damage from potential cyber threats.

PART 4

NAVIGATING REGULATORY COMPLIANCE IN FINTECH SECURITY

CHAPTER 9
KEY CYBERSECURITY REGULATIONS

📖 Introduction

The rapid evolution of financial technology (FinTech) has transformed how businesses and consumers interact with financial services. However, this digital revolution comes with significant cybersecurity risks, prompting global regulatory bodies to establish stringent cybersecurity frameworks. This chapter delves into key cybersecurity regulations including the Payment Card Industry Data Security Standard (PCI DSS), the General Data Protection Regulation (GDPR)[158], and the Revised Payment Services Directive (PSD2). It also covers practical steps for compliance, integration of these regulations into the product development lifecycle, and lessons from real-world cases of regulatory breaches.

📖 Section 1: Payment Card Industry Data Security Standard (PCI DSS)

Overview

PCI DSS is a globally recognized security standard created to ensure the safe handling of credit card information. It was established by

---

[158]https://www.pcisecuritystandards.org/document_library/

major credit card brands such as Visa, MasterCard, American Express, and Discover.

Core Requirements

The PCI DSS framework consists of 12 key requirements divided into six control objectives:

Build and Maintain a Secure Network and Systems:

- **Install and maintain a firewall configuration to protect cardholder data:** Implement a robust firewall system that monitors incoming and outgoing traffic, ensuring that only authorized communications are allowed. Regularly update firewall configurations to reflect evolving security requirements and emerging threats.

- **Avoid using vendor-supplied defaults for system passwords and other security settings:** Change default passwords, usernames, and other system credentials immediately after installation. Configure security settings based on organizational policies, ensuring they align with industry best practices.

Protect Cardholder Data:

- **Protect stored cardholder data through encryption:** Use strong encryption algorithms to secure stored cardholder information. Ensure encryption keys are securely managed, rotated periodically, and restricted to authorized personnel only.

- **Use strong encryption for transmission over open, public networks:** Secure sensitive data transmitted over public networks using protocols such as TLS (Transport Layer

Security) or IPSec. Regularly update encryption protocols to comply with the latest security standards.

Maintain a Vulnerability Management Program:

- **Use regularly updated anti-virus software:** Install and maintain anti-virus and anti-malware software on all devices that interact with sensitive data. Schedule automatic updates and frequent system scans to detect and remove potential threats promptly.

- **Develop and maintain secure systems and applications:** Ensure all systems, software, and applications are updated with the latest security patches. Conduct regular code reviews, vulnerability assessments, and penetration testing to identify and mitigate security risks.

Implement Strong Access Control Measures:

- **Restrict access to cardholder data on a need-to-know basis:** Apply the principle of least privilege, granting access only to employees who require it for legitimate business purposes. Use role-based access controls to limit the scope of data exposure.

- **Identify and authenticate access to system components:** Implement multi-factor authentication (MFA) and strong password policies to verify the identity of users accessing sensitive systems. Regularly review and update access permissions.

- **Restrict physical access to cardholder data:** Secure physical locations containing sensitive data using access control mechanisms such as keycards, biometric scanners, and surveillance cameras. Maintain a visitor log to track who enters secure areas.

Regularly Monitor and Test Networks:

- **Track and monitor all access to network resources and cardholder data:** Implement comprehensive logging and monitoring tools to track all system access attempts. Review logs regularly for unusual activities or security incidents, ensuring prompt investigation and response.

- **Conduct regular security testing:** Schedule periodic security assessments, including vulnerability scanning and penetration testing. Use the results to improve system defenses and close potential security gaps.

Maintain an Information Security Policy:

- **Establish a security policy addressing information security for employees and contractors:** Develop a clear and comprehensive information security policy outlining roles, responsibilities, and acceptable use of organizational resources. Conduct regular security awareness training for all employees and contractors to ensure they understand their obligations and best practices for maintaining data security.

- **Ensure continuous policy reviews and updates:** Regularly review and update the security policy to reflect changes in technology, business operations, and emerging security threats. Communicate policy updates to all relevant stakeholders and enforce compliance through audits and assessments.

Compliance Levels

Merchants and service providers are categorized into levels based on transaction volume, determining the required validation procedures.

Challenges and Best Practices

- Regular security assessments
- Continuous employee training
- Use of advanced security technologies

📖 Section 2: General Data Protection Regulation (GDPR)

Overview

The GDPR, enacted by the European Union (EU), governs the collection, processing, and storage of personal data. It applies to organizations operating within the EU or handling EU citizens' data.

Key Principles

Lawfulness, Fairness, and Transparency:

- **Lawfulness:** Organizations must ensure that personal data is processed only when there is a legal basis, such as consent from the data subject, contractual obligations, legal requirements, protection of vital interests, or legitimate interests pursued by the data controller or third party.

- **Fairness:** Data processing must be conducted in a way that respects the rights and expectations of data subjects. This includes avoiding deceptive practices and ensuring individuals are not subjected to harmful or discriminatory processing activities.

- **Transparency:** Data subjects must be clearly informed about how their data is being used. This includes providing accessible privacy notices detailing what data is collected, why it is collected, how long it will be stored, and with whom it may be shared.

Purpose Limitation:

- Personal data should only be collected for specific, explicit, and legitimate purposes. Organizations must clearly define the purpose of data collection before starting the process.

- Data must not be processed in ways incompatible with the original purposes unless further consent is obtained or legal grounds justify additional processing.

- This principle ensures that organizations cannot repurpose data without informing data subjects and obtaining appropriate legal permissions.

Data Minimization:

- Only the minimum amount of personal data necessary for the specified purpose should be collected.

- Organizations must assess and determine the essential data required to achieve their business goals while minimizing the collection of excessive or irrelevant information.

- Regular reviews should be conducted to ensure unnecessary data is not collected or retained.

Accuracy:

- Personal data must be kept accurate, complete, and up-to-date. Inaccurate or incomplete data should be corrected or deleted promptly.

- Organizations should implement mechanisms for data subjects to request corrections to their personal information when necessary.

- Automated processes should be regularly audited to prevent incorrect data processing.

Storage Limitation:

- Data should only be stored for as long as necessary to fulfill the purposes for which it was collected. After this period, personal data must be securely deleted or anonymized.

- Organizations must define clear data retention policies that specify how long various types of personal data will be retained.

- Compliance reviews should ensure that data retention periods are adhered to, and expired data is securely disposed of.

Integrity and Confidentiality:

- Personal data must be protected against unauthorized access, disclosure, alteration, or destruction.

- Security measures such as data encryption, access controls, secure storage, and regular vulnerability assessments should be implemented.

- Organizations must conduct regular security training for employees and maintain an incident response plan to address potential breaches or data leaks.

Accountability:

- Organizations must take responsibility for complying with data protection laws and demonstrating their compliance through documented policies, procedures, and practices.

- Data protection officers (DPOs) or similar roles should be appointed to oversee data privacy compliance efforts.

- Regular audits, impact assessments, and compliance reports should be conducted to identify risks and improve data management practices.

- Organizations should be prepared to cooperate with regulatory authorities and respond to data subject requests in a timely and transparent manner.

Data Subject Rights

- Right to Access
- Right to Rectification
- Right to Erasure ("Right to be Forgotten")
- Right to Data Portability
- Right to Object

Compliance Requirements

- Conduct data protection impact assessments (DPIAs).
- Appoint a Data Protection Officer (DPO) if necessary.
- Report data breaches within 72 hours.

Penalties for Non-Compliance

Fines can reach up to 4% of global annual turnover or €20 million, whichever is higher.

📖 Section 3: Revised Payment Services Directive (PSD2)

Overview

PSD2[159] is a European Union directive aimed at creating safer and more innovative payment services within the EU. It enhances customer protection and promotes competition by requiring strong customer authentication (SCA) and secure communication.

Core Requirements

Strong Customer Authentication (SCA):

1. **Definition:** Strong Customer Authentication (SCA) is a security measure required for electronic payments to reduce fraud and enhance payment security. It mandates the use of multi-factor authentication (MFA).

2. **Authentication Factors:** SCA requires the use of at least two of the following three authentication elements:

    o   **Something the user knows:** Examples include a password, PIN, or answer to a security question.

    o   **Something the user has:** Examples include a smartphone, token, or smart card.

    o   **Something the user is:** Examples include biometric identifiers such as fingerprints, facial recognition, or voice patterns.

---

[159]https://www.ecb.europa.eu/press/intro/mip-online/2018/html/1803_revisedpsd.en.html

3. **Applicability:** SCA applies to most online payments and access to sensitive account information within the European Economic Area (EEA), ensuring secure digital transactions.

Access to Account (XS2A):

1. **Definition:** Access to Account (XS2A) is a regulatory requirement under the Revised Payment Services Directive (PSD2) that mandates banks to grant third-party providers (TPPs) access to customer accounts with the customer's explicit consent.

2. **How It Works:**

   o Banks must create secure APIs (Application Programming Interfaces) to allow TPPs to access customer account information and initiate payments on the customer's behalf.

   o Customers retain control by granting or revoking consent at any time.

3. **Types of TPPs:**

   o **Account Information Service Providers (AISPs):** These providers aggregate customer account data for financial management or advisory purposes.

   o **Payment Initiation Service Providers (PISPs):** These providers initiate payments directly from customer accounts, simplifying the payment process.

4. **Security and Compliance:** Banks and TPPs must comply with strict security standards, ensuring customer data is securely transmitted and processed.

Transparency and Consumer Protection:

1. **Clear and Detailed Disclosure of Fees:**

- o Payment service providers (PSPs) must disclose all fees, charges, and terms in a clear, concise, and transparent manner before a customer initiates a transaction.
- o This includes showing exchange rates, service fees, and transaction costs to avoid hidden charges.

2. **Real-Time Transaction Notifications:**

- o Customers must receive immediate notifications for transactions involving their accounts, such as payments, withdrawals, or fund transfers.
- o Notifications can be sent via SMS, email, or mobile apps, enhancing financial control and fraud prevention.

3. **Consumer Rights Protections:**

- o Customers are entitled to dispute unauthorized or incorrect transactions, with financial institutions required to investigate and resolve complaints within defined timeframes.
- o Regulatory frameworks ensure customers are protected from fraudulent activities, unauthorized access, and data breaches.

Open Banking:

1. **Definition:** Open Banking refers to the practice of enabling secure data sharing between banks and third-party developers through APIs, fostering financial innovation and customer-centric services.

2. **How It Works:**

- o Banks provide secure, regulated access to customer account data when customers authorize third-party applications or services.

    o Developers can create innovative financial products, such as budgeting apps, investment platforms, or loan comparison tools.

3. **Benefits:**

    o **Enhanced Financial Services:** Customers gain access to a wider range of personalized financial services, improving convenience and choice.

    o **Competition and Innovation:** Open Banking drives competition among financial service providers, reducing costs and enhancing service quality.

    o **Customer Control:** Customers retain control over who can access their financial data and for what purposes, ensuring privacy and data security.

4. **Regulatory Frameworks:**

    o Open Banking is governed by regulations like PSD2 in the EU and similar frameworks globally. These regulations ensure that data sharing occurs securely, transparently, and with the customer's full consent.

Compliance Steps

- Implement multi-factor authentication (MFA).
- Secure application programming interfaces (APIs).
- Monitor and respond to security incidents proactively.

📖 Section 4: Practical Steps for Compliance

Regulatory Gap Analysis:

1. **Conduct a Comprehensive Assessment of Current Cybersecurity Measures:**

   - Perform a thorough review of existing cybersecurity frameworks, policies, and technologies in place.
   - Evaluate the effectiveness of current security controls, including firewalls, intrusion detection systems (IDS), and access management solutions.
   - Document and analyze current data protection practices across all business functions.

2. **Identify Gaps in Compliance with PCI DSS, GDPR, and PSD2:**

   - Compare existing security measures against specific compliance requirements for regulations such as:

     - **PCI DSS:** Focus on protecting payment card information.

     - **GDPR:** Address personal data privacy and consent management.

     - **PSD2:** Ensure secure electronic payments and strong customer authentication (SCA).

   - Highlight deficiencies, such as missing security controls, outdated policies, or incomplete documentation.

Develop a Compliance Roadmap:

   - **Create an Action Plan with Timelines and Responsibilities:**

- Develop a detailed action plan outlining specific tasks, milestones, and deadlines for closing identified compliance gaps.

- Assign roles and responsibilities to internal teams, ensuring accountability for each compliance task.

- Prioritize actions based on risk level, focusing first on critical vulnerabilities that could lead to data breaches or non-compliance penalties.

- **Resource Allocation and Budget Planning:**

  - Allocate appropriate budgets for system upgrades, training programs, and additional staff if required.

  - Ensure senior management support for the compliance initiative to facilitate decision-making and resource mobilization.

Security Controls Implementation:

- **Implement Firewalls, Encryption Protocols, and MFA:**

  - Install next-generation firewalls (NGFW) to secure network perimeters and filter incoming and outgoing traffic.

  - Apply industry-standard encryption protocols such as TLS and AES to protect sensitive data at rest and in transit.

  - Enforce multi-factor authentication (MFA) for all critical systems and services to strengthen access control.

- **Conduct Regular Penetration Testing:**

  - Schedule external and internal penetration testing to identify system vulnerabilities.

  - Simulate potential attacks on network infrastructures, applications, and APIs to uncover weaknesses.

   o Use test results to enhance security controls and ensure continuous system improvement.

Training and Awareness:

- **Conduct Regular Training Sessions for Employees:**

  o Organize mandatory cybersecurity training programs for employees at all organizational levels.

  o Cover topics such as phishing prevention, password management, and secure data handling practices.

  o Provide role-specific training for employees in sensitive areas such as IT, finance, and human resources.

- **Promote a Culture of Cybersecurity Awareness:**

  o Launch ongoing awareness campaigns using newsletters, posters, and online resources.

  o Celebrate Cybersecurity Awareness Month by hosting events, workshops, and interactive sessions.

  o Encourage employees to report potential threats or suspicious activities without fear of repercussions.

Regular Audits and Monitoring:

- **Schedule Routine Internal and External Audits:**

  o Perform regular internal audits to assess compliance with regulatory standards and organizational policies.

  o Hire certified third-party auditors for independent assessments and certifications when required.

  o Document findings and create corrective action plans based on audit recommendations.

- **Monitor Systems for Suspicious Activities:**
  - Implement advanced Security Information and Event Management (SIEM) tools to monitor network activity in real time.
  - Set up automated alerts for unusual patterns such as unauthorized access attempts, data transfers, and privilege escalations.
  - Conduct continuous log monitoring and periodic reviews of system activity to detect and respond to potential breaches.

Incident Response Plan:

- **Establish Clear Protocols for Responding to Data Breaches:**
  - Develop a comprehensive incident response (IR) policy detailing roles, responsibilities, and reporting procedures.
  - Define clear incident severity levels and appropriate response actions for each level.
  - Ensure that contact information for incident response team members and external partners is up-to-date.

- **Conduct Regular Incident Response Drills:**
  - Schedule regular mock breach simulations to test the organization's incident response readiness.
  - Include cross-functional teams such as IT, legal, public relations, and management in simulations.
  - Conduct post-mortem analyses after each drill to identify areas for improvement and refine the incident response plan accordingly.

- **Communication and Reporting:**
  - o Establish internal and external communication protocols, ensuring timely notifications to affected individuals, regulators, and stakeholders in the event of a breach.
  - o Maintain detailed incident records for future reference, compliance audits, and legal purposes.

📖 Section 5: Case Studies and Lessons Learned

Case Study 1: PCI DSS Non-Compliance

A global retail chain experienced a massive data breach due to a combination of poor firewall configuration and outdated anti-virus software. Cybercriminals exploited these vulnerabilities, gaining unauthorized access to the company's payment processing systems. As a result, millions of credit card records were exposed, causing severe financial and reputational damage. The breach led to significant fines imposed by regulatory bodies, costly legal battles, and a steep decline in customer trust. Additionally, the company had to invest heavily in upgrading its cybersecurity infrastructure and retraining its staff[160].

Lesson: Regular security audits, timely system updates, and continuous employee training are crucial to maintaining PCI DSS compliance and protecting sensitive customer information. Companies must adopt a proactive approach to cybersecurity by conducting regular penetration tests, patching known

---

[160] https://www.securitymetrics.com/static/resources/orange/2017-securitymetrics-pci-guide.pdf

vulnerabilities, and monitoring network traffic for suspicious activities[161].

Case Study 2: GDPR Breach

A major tech company was fined €50 million for violating the General Data Protection Regulation (GDPR). The investigation revealed that the company failed to obtain proper user consent for data collection and processing practices. Personal data was gathered without clear communication about how it would be used, stored, or shared with third parties. Furthermore, the company's privacy policy was complex and difficult for users to understand, exacerbating the compliance failure.

The fine not only impacted the company's financial standing but also triggered a wave of negative media coverage and a loss of consumer trust. To regain credibility, the company had to overhaul its data protection policies, simplify its privacy terms, and implement a transparent user consent process[162].

Lesson: Transparent data collection policies, clear user consent mechanisms, and straightforward privacy notices are essential for GDPR compliance. Organizations should regularly review and update their data privacy policies, ensuring they align with current regulations and are easily understood by users. Employing privacy-by-design principles and appointing a dedicated Data Protection Officer (DPO) can also strengthen compliance efforts[163].

---

[161] https://www.securitymetrics.com/static/resources/orange/2017-securitymetrics-pci-guide.pdf

[162] https://www.dataprotectionreport.com/2019/01/first-multi-million-euro-gdpr-fine-google-llc-fined-e50-million-under-gdpr-for-transparency-and-consent-infringements-in-relation-to-use-of-personal-data-for-personalised-ads-2/

[163] https://termly.io/resources/articles/biggest-gdpr-fines/

Case Study 3: PSD2 Violation

A financial services firm faced severe penalties after failing to implement Strong Customer Authentication (SCA), a key requirement under the Payment Services Directive 2 (PSD2). The oversight left the company vulnerable to fraudulent transactions, resulting in significant monetary losses and regulatory fines. Customers reported unauthorized transactions, leading to numerous complaints and claims for compensation.

The company's failure to adopt robust authentication measures not only caused financial damage but also tarnished its reputation. To recover, the firm had to deploy a comprehensive fraud prevention system, introduce two-factor authentication, and strengthen its customer verification processes[164].

Lesson: Strong authentication protocols, continuous monitoring, and adherence to PSD2 requirements are essential to prevent fraud and ensure regulatory compliance. Financial institutions should invest in advanced fraud detection technologies, perform regular risk assessments, and provide customer education on safe banking practices. Collaboration with cybersecurity experts can further enhance an organization's ability to respond swiftly to emerging threats.

164

https://www.eba.europa.eu/sites/default/files/document_library/Publications/Opinions/2022/Opinion%20od%20PSD2%20review%20%28EBA-Op-2022-06%29/1036016/EBA%27s%20response%20to%20the%20Call%20for%20advice%20on%20the%20review%20of%20PSD2.pdf

📖 Conclusion

Adhering to cybersecurity regulations like PCI DSS, GDPR, and PSD2 is critical for FinTech companies. These frameworks help safeguard sensitive data, build customer trust, and avoid severe penalties. A proactive approach that includes robust security practices, regular training, and compliance monitoring ensures a resilient and secure digital financial ecosystem.

💡 What We Learnt:

- **Cybersecurity Regulations Are Critical:** Adhering to frameworks like PCI DSS, GDPR, and PSD2 ensures the protection of sensitive data, builds customer trust, and avoids legal penalties.

- **PCI DSS Compliance Safeguards Payment Data:** Organizations handling payment data must implement strong firewalls, encryption protocols, multi-factor authentication, and continuous monitoring to meet PCI DSS standards.

- **GDPR Protects Personal Data Privacy:** Businesses must ensure data collection is lawful, transparent, and secure, while respecting individual rights such as access, correction, and data deletion.

- **PSD2 Enhances Payment Security:** Strong Customer Authentication (SCA), secure APIs, and open banking foster safer, more competitive digital payment environments.

- **Compliance Requires a Multi-Layered Approach:** Companies should perform regulatory gap analyses, develop compliance roadmaps, and allocate adequate resources for security controls and staff training.

- **Training and Awareness Are Essential:** Regular employee training on cybersecurity best practices helps prevent human errors that could lead to data breaches.

- **Audits and Monitoring Ensure Ongoing Security:** Internal and external audits, real-time monitoring, and penetration testing help identify vulnerabilities and strengthen cybersecurity defenses.

- **Incident Response Plans Mitigate Breaches:** Having a clear incident response strategy enables organizations to react quickly and effectively to security breaches, minimizing damage and maintaining regulatory compliance.

- **Case Studies Highlight Real-World Consequences:** Learning from industry breaches underscores the importance of compliance, proactive monitoring, and transparent data practices.

- **Continuous Improvement Is Necessary:** Regular updates to security policies, technologies, and processes ensure resilience against evolving cyber threats and changing regulatory requirements.

CHAPTER 10
MANAGING COMPLIANCE THROUGHOUT THE PRODUCT LIFECYCLE

📖 Building a Compliance-Driven Culture

A compliance-driven culture[165] ensures that regulatory adherence becomes a natural part of everyday operations. Organizations that embed compliance into their culture not only avoid costly penalties but also enhance their reputation and operational efficiency.

Establishing Policies and Practices That Prioritize Compliance

Building a culture centered on compliance requires organizations to establish and enforce clear policies and practices. A compliance-driven environment not only mitigates risks but also enhances an organization's reputation, operational efficiency, and long-term sustainability. Below are key strategies for fostering such a culture:

Preparing for Audits and Handling Reporting Obligations

Regular audits and transparent reporting are critical components of a robust compliance strategy. Audits help ensure that business processes align with legal and regulatory requirements, while

---

[165] https://www.diligent.com/resources/blog/culture-of-compliance

accurate reporting builds trust with stakeholders and regulatory agencies.

- **Internal Audits:** Conducting periodic internal audits helps identify potential compliance gaps before external audits occur.

- **External Audits:** Collaborating with third-party auditors brings an unbiased perspective and ensures adherence to global standards.

- **Accurate and Timely Reporting:** Providing timely and truthful reports on compliance-related matters fosters credibility and transparency.

- **Data Management:** Maintaining secure and organized records simplifies the reporting process and supports compliance verification.

By implementing these strategies, organizations can build a resilient compliance framework that mitigates risks, enhances corporate reputation, and drives sustainable success.

Establishing Policies and Practices That Prioritize Compliance

Building a culture centered on compliance requires organizations to establish and enforce clear policies and practices. A compliance-driven environment not only mitigates risks but also enhances an organization's reputation, operational efficiency, and long-term sustainability. Below are key strategies for fostering such a culture:

1. Define Clear Compliance Objectives

Creating an effective compliance framework[166] begins with defining clear compliance objectives that align with an organization's overarching business goals. This alignment ensures that compliance is not viewed as a separate or burdensome function but rather as an integral part of daily business operations, driving efficiency and reducing risk. When compliance goals are directly linked to business objectives, companies can more easily foster a culture of accountability and transparency.

Key Steps to Defining Compliance Objectives:

Understand Industry-Specific Regulations:

Organizations must stay well-informed about the regulatory environment relevant to their industry. This involves conducting regular reviews of current laws, industry standards, and best practices. Compliance teams should collaborate with legal experts and industry associations to interpret complex regulations accurately. Additionally, companies should consider conducting periodic audits and risk assessments to identify potential compliance gaps and ensure ongoing adherence to applicable laws.

Communicate Compliance Requirements:

Effective communication is the cornerstone of a successful compliance program. Organizations should ensure that compliance policies and procedures are clearly articulated and easily accessible to all employees. This can be achieved by incorporating compliance expectations into employee handbooks, updating company intranets regularly, and conducting mandatory compliance training sessions. Regular briefings and internal memos can further reinforce the importance of compliance, while an open-door policy can encourage

---

[166]https://www.resolver.com/resource/how-to-build-compliance-culture/

employees to raise concerns or seek clarification without fear of retaliation.

Set Measurable Goals:

Establishing clear, measurable, and actionable compliance goals allows organizations to track progress and evaluate the effectiveness of their compliance initiatives. These goals should follow the SMART framework:

- o **Specific:** Clearly define what needs to be achieved.

- o **Measurable:** Set quantifiable indicators of progress.

- o **Achievable:** Ensure the goals are realistic given available resources.

- o **Relevant:** Align goals with broader business and compliance priorities.

- o **Time-bound:** Assign deadlines to drive timely completion.

   For example, a company might set a goal to complete annual compliance training for all employees by a specific date or reduce incidents of regulatory non-compliance by a certain percentage within a fiscal year.

Additional Considerations:

Develop Clear Policies and Procedures: Documenting compliance policies and procedures ensures consistency and transparency in how compliance tasks are carried out.

Assign Accountability: Clearly define roles and responsibilities related to compliance to avoid ambiguity and ensure ownership.

Monitor and Review: Regularly monitor compliance performance through audits, inspections, and feedback channels. Adjust goals and strategies as needed based on findings.

Foster a Compliance Culture: Promote a culture where compliance is valued and integrated into the company's core values, ensuring long-term sustainability and reduced regulatory risk.

By defining clear compliance objectives through these steps, organizations can build a robust compliance framework that not only minimizes legal risks but also enhances operational efficiency and stakeholder confidence.

2. Leadership Commitment

A compliance-driven culture begins with strong leadership. Senior leaders and executives set the tone for the entire organization by defining and reinforcing its core values and ethical standards. Their commitment to compliance and integrity serves as a foundation for building a trustworthy and transparent workplace. Leadership involvement in compliance initiatives demonstrates that ethical conduct is not merely a policy but a strategic priority that guides decision-making at every level.

Key Ways Leaders Can Foster a Compliance-Driven Culture:

Modeling Ethical Behavior:

Leaders must lead by example. Their actions should consistently reflect the organization's commitment to ethical conduct, honesty, and compliance with applicable laws and regulations. When leaders visibly adhere to compliance standards, employees are more likely to follow suit, understanding that ethical behavior is expected and rewarded.

Supporting Compliance Programs:

Senior management should actively support compliance initiatives by allocating adequate financial, technological, and human resources. This support ensures that compliance programs are well-equipped to operate effectively. Leaders should participate in compliance-related training sessions to demonstrate their personal investment in the organization's ethical development.

Establishing Clear Policies and Procedures:
Leaders play a pivotal role in developing and maintaining clear, well-documented policies and procedures. By ensuring these guidelines are accessible, understandable, and regularly updated, they help create a structured environment where employees know their responsibilities and the consequences of non-compliance.

Communicating the Importance of Compliance:
Regular, transparent communication from senior leadership reinforces the significance of compliance and ethical behavior. This includes delivering messages through various channels such as company-wide meetings, emails, newsletters, and even informal discussions. Frequent and clear communication highlights that compliance is not just a regulatory requirement but a core component of the company's mission and values.

Encouraging Open Dialogue and Reporting:
Leaders should cultivate an environment where employees feel safe reporting concerns about unethical behavior or potential compliance violations. Establishing anonymous reporting channels and guaranteeing protection against retaliation encourages openness. Regularly addressing reported issues shows that leadership takes compliance concerns seriously.

Recognizing and Rewarding Ethical Behavior:
Recognizing employees who demonstrate integrity and compliance reinforces positive behavior. Leaders can introduce rewards, public recognition, or other incentives to promote a culture where doing the right thing is valued and celebrated.

Evaluating and Improving Compliance Efforts:
Leadership should regularly review the effectiveness of compliance programs through audits, employee feedback, and performance evaluations. Continuous improvement ensures that the compliance framework evolves with regulatory changes and organizational growth.

By embedding these practices into their daily actions, senior leaders create a culture of compliance that permeates the entire organization. When employees see that leadership is genuinely committed to ethical behavior, they are more likely to adopt similar values, fostering a workplace where integrity and accountability[167] thrive.

3. Training and Education

Employees must be well-informed about compliance policies and legal obligations relevant to their roles. A knowledgeable workforce is critical for maintaining a culture of integrity and accountability within an organization. Regular training programs ensure that staff members understand how to meet compliance standards effectively and reduce potential risks associated with non-compliance. Key components of an effective training and education program include:

1. Mandatory Onboarding Training
Introducing new hires to key compliance requirements during the onboarding process helps close knowledge gaps early on. This foundational training covers essential policies, regulatory expectations, and industry-specific compliance standards. By familiarizing new employees with these critical areas from the beginning, organizations set a precedent for responsible conduct and informed decision-making.

---

[167]https://eimf.eu/the-importance-of-a-strong-compliance-culture-in-organisations/

2. Interactive Workshops

Engaging workshops and role-playing scenarios play a vital role in helping employees grasp complex regulatory concepts. These sessions provide hands-on experience through simulated real-world situations, enhancing both understanding and retention. Interactive workshops encourage active participation, foster collaboration, and promote problem-solving skills critical for compliance-related decision-making.

3. Online Courses and Certifications

Offering accessible online modules enables continuous learning and compliance certification. These courses can be tailored to specific roles and updated regularly to reflect the latest regulatory changes. Employees can complete modules at their own pace, ensuring flexibility while maintaining high training standards. Certification programs also serve as formal recognition of employees' commitment to compliance excellence.

4. Ongoing Refresher Sessions

Periodic refresher sessions are essential to reinforce key compliance principles and update employees on emerging regulatory trends. These sessions help prevent knowledge decay and ensure that staff remains current on best practices. Refresher courses can be conducted through webinars, workshops, or targeted briefings focusing on recent policy updates.

5. Compliance Communication Channels

Establishing clear communication channels ensures that employees have access to expert guidance when needed. Dedicated compliance hotlines, email support, and intranet resources can provide timely assistance and clarification. Encouraging an open-door policy where employees can seek advice without fear of reprisal fosters a transparent and supportive compliance culture.

6. Performance Tracking and Feedback

Monitoring training completion rates and assessing employee understanding through quizzes, tests, and feedback surveys help measure the effectiveness of training programs. Regular performance evaluations enable organizations to identify knowledge gaps and adjust training content accordingly. Constructive feedback also motivates employees to stay proactive in their learning journey.

In summary, a comprehensive training and education framework is indispensable for fostering a culture of compliance and reducing legal risks. By investing in continuous learning initiatives, organizations can ensure that employees are well-equipped to navigate regulatory complexities while upholding ethical standards.

3. *Employee Accountability*

Accountability at all organizational levels is essential for reinforcing a culture of compliance within a company. When employees understand that they are responsible for their actions and decisions, they are more likely to follow established rules and guidelines. Companies can achieve this through a variety of strategies designed to promote transparency, responsibility, and adherence to organizational policies.

Key Strategies for Promoting Accountability

1. **Performance Reviews:** Regular performance evaluations that include compliance metrics ensure that employees are held accountable for adhering to company policies. Incorporating these metrics into annual or quarterly reviews makes compliance an integral part of job performance. Managers should provide constructive feedback, highlighting areas where employees excel in compliance and identifying opportunities for improvement. This approach

creates a continuous feedback loop that supports ongoing development.

2. **Rewards and Recognition:** Recognizing and rewarding employees who consistently demonstrate compliance-minded behavior encourages others to follow suit. Rewards can be both monetary, such as bonuses or gift cards, and non-monetary, like public recognition during meetings or through internal communications. Creating a culture where compliance is celebrated helps reinforce its importance as a core company value.

3. **Enforcement of Consequences:** Establishing clear disciplinary procedures for instances of non-compliance serves as a powerful deterrent against misconduct. Employees should be aware of the potential repercussions of violating company policies, including verbal warnings, formal reprimands, suspension, or termination. Transparent communication about these procedures ensures that all employees understand the stakes involved.

Case Study: PharmaTech Inc.

PharmaTech Inc., a global pharmaceutical company[168], implemented a compliance-first strategy by embedding regulatory training into its onboarding program. Recognizing the complexities of operating in a highly regulated industry, the company designed a comprehensive 60-day intensive training course for all new hires. This program covers critical areas such as international industry regulations, data privacy laws, clinical trial standards, and manufacturing protocols.

---

[168] https://pharma-tech.com/

The training program includes interactive modules, scenario-based learning, and assessments to ensure that employees grasp key compliance principles. PharmaTech also introduced periodic refresher courses and required certifications to maintain high compliance standards throughout employees' careers.

As a result of these initiatives, PharmaTech experienced a 40% reduction in compliance breaches within a year. This significant improvement underscored the tangible impact of a robust compliance training program, reinforcing the importance of proactive employee education. Additionally, the company's reputation among regulatory bodies improved, leading to faster approvals for new products and smoother regulatory inspections.

Employee accountability is a cornerstone of a successful compliance culture. By integrating compliance metrics into performance reviews, recognizing positive behavior, and enforcing clear disciplinary measures, companies can build a resilient and responsible workforce. PharmaTech Inc.'s experience demonstrates that a strategic focus on training and accountability can lead to measurable improvements, fostering long-term organizational success and integrity[169].

📖 Preparing for Audits and Handling Reporting Obligations

Regular audits and transparent reporting are critical components of a robust compliance strategy. Audits help ensure that business processes align with legal and regulatory requirements, while

---

[169] https://blog.rewardian.com/5-effective-onboarding-training-strategies-for-pharmaceutical-companies

accurate reporting builds trust with stakeholders and regulatory agencies.

- **Internal Audits:** Conducting periodic internal audits helps identify potential compliance gaps before external audits occur.

- **External Audits:** Collaborating with third-party auditors brings an unbiased perspective and ensures adherence to global standards.

- **Accurate and Timely Reporting:** Providing timely and truthful reports on compliance-related matters fosters credibility and transparency.

- **Data Management:** Maintaining secure and organized records simplifies the reporting process and supports compliance verification.

By implementing these strategies, organizations can build a resilient compliance framework that mitigates risks, enhances corporate reputation, and drives sustainable success.

1. Tools and Techniques for Managing Audits and Maintaining Transparent Records

- **Audit Preparation:** Conduct internal audits regularly to identify gaps before external inspections. Use audit checklists tailored to industry standards.

- **Documentation Management:** Maintain well-organized records. Cloud-based solutions like compliance management software can streamline data storage and retrieval.

- **Real-Time Monitoring:** Implement monitoring tools that track compliance metrics. Automated alerts can flag potential issues, allowing teams to address them proactively.

Case Study: GreenTech Solutions

GreenTech Solutions, a renewable energy company, faced regulatory scrutiny after inconsistent audit reports. After implementing a centralized compliance management system, they achieved a 98% compliance rate within two reporting cycles, earning industry recognition for transparency[170].

Tips for Implementing Regulatory Updates Efficiently

Staying ahead of regulatory changes is essential for maintaining compliance. Delayed implementation can result in operational disruptions and financial penalties.

1. Adapting to Changes in Regulation and Incorporating Updates into Product Roadmaps

- **Regulatory Monitoring:** Assign a dedicated team or use automated services to track changes in relevant regulations.

- **Impact Assessment:** Evaluate how new regulations affect existing products and processes. Conduct risk assessments to prioritize actions.

- **Change Management:** Develop a clear process for updating product development roadmaps. Communicate changes across all relevant teams to ensure alignment.

- **Stakeholder Engagement:** Involve key stakeholders, including suppliers and clients, when implementing regulatory changes to avoid misunderstandings.

---

[170] https://greentech.energy/wp-content/uploads/2022_greentech_SustainabilityReport-2.pdf

Case Study: MedEquip Manufacturing

MedEquip Manufacturing successfully navigated a complex regulatory overhaul in the medical device industry by establishing a regulatory task force. The team provided monthly updates, ensuring that all product lines met new compliance standards six months ahead of the deadline[171].

📖 Conclusion

Managing compliance throughout the product lifecycle requires a multi-faceted approach that integrates policy development, regular audits, and proactive adaptation to regulatory changes. By fostering a compliance-driven culture, using modern auditing tools, and staying updated on industry standards, organizations can minimize risks and maintain a strong market presence. The case studies of PharmaTech Inc., GreenTech Solutions, and MedEquip Manufacturing highlight how strategic compliance management leads to sustainable success.

💡 What We Learnt:

- **Building a Compliance-Driven Culture:** Organizations must embed compliance into their culture to avoid penalties, enhance their reputation, and improve operational efficiency.

---

[171]

http://ndl.ethernet.edu.et/bitstream/123456789/91873/1/Health%20Care%20Administration_%20Managing%20Organized%20Delivery%20Systems%2C%20Fifth%20Edition%20%28Health%20Care%20Administration%20%28Wolper%29%29%20%28%20PDFDrive%20%29.pdf

- **Establishing Policies and Practices:** Clear policies, well-defined roles, and consistent procedures ensure compliance becomes a seamless part of business operations.

- **Defining Compliance Objectives:** Compliance goals should align with business objectives, be specific, measurable, achievable, relevant, and time-bound (SMART).

- **Leadership Commitment:** Senior management must model ethical behavior, allocate resources, and communicate the importance of compliance consistently.

- **Training and Education:** Regular training through onboarding sessions, interactive workshops, online courses, and refresher programs keeps employees informed and compliant.

- **Employee Accountability:** Performance reviews, recognition programs, and clear disciplinary procedures promote responsibility and adherence to compliance standards.

- **Preparing for Audits and Reporting Obligations:** Conducting internal and external audits, ensuring accurate reporting, and maintaining organized records strengthen transparency and regulatory compliance.

- **Adapting to Regulatory Changes:** Organizations must monitor regulations, assess their impact, and adjust their product development roadmaps accordingly.

- **Tools and Techniques for Audits:** Use audit preparation checklists, documentation management systems, and real-time monitoring tools for efficient compliance tracking.

- **Case Studies as Examples:** Real-world examples from companies like PharmaTech Inc., GreenTech Solutions, and MedEquip Manufacturing demonstrate how strategic compliance management can drive sustainable success.

PART 5

THE ROLE OF THE PRODUCT MANAGER IN CYBERSECURITY

CHAPTER 11
BALANCING SECURITY AND USABILITY

The rise of FinTech has revolutionized financial services but has also magnified the challenge of balancing cybersecurity with product usability. Product managers play a pivotal role in navigating this balance to ensure a seamless user experience while maintaining robust security protocols. This chapter explores how product managers can integrate security into product development, align it with core business goals, advocate for resources, and educate stakeholders through detailed case studies and actionable strategies.

📖 The Product Manager's Role in Cybersecurity

Product managers operate at the intersection of business strategy, customer needs, and technical execution. Their role is multifaceted, involving collaboration across various teams to ensure products align with business goals while meeting user expectations. A critical, often underemphasized, aspect of this responsibility is cybersecurity. As digital threats evolve, product managers must integrate security considerations into every stage of product development, from ideation to release and beyond.

Key Responsibilities of Product Managers in Cybersecurity

Prioritizing Security Features:

Product managers must strike a delicate balance between implementing robust security measures and maintaining product usability. Security features like biometric authentication, end-to-end encryption, secure payment gateways, and data masking enhance product safety but can also add complexity to the user experience. A successful product manager ensures that security is built into the product without compromising ease of use. They work closely with UX designers and developers to create seamless, secure experiences.

Product managers also need to evaluate emerging technologies such as zero-trust frameworks, decentralized identity management, and artificial intelligence-driven security tools. Integrating these technologies can provide competitive advantages while enhancing overall system protection.

Stakeholder Alignment:

Product managers serve as the bridge between technical teams, business executives, marketing departments, and customers. They must communicate the importance of cybersecurity to all stakeholders. This includes advocating for necessary security investments, explaining the potential risks of security breaches, and ensuring that the technical team understands business priorities related to security.

Regular security reviews, threat assessments, and incident response planning should be part of the product management process. Product managers should organize cross-functional meetings to discuss security updates and ensure that everyone is aligned on roles and responsibilities in maintaining product security.

Security-First Mindset:

Embedding security into the product roadmap from the beginning is essential. This proactive approach reduces the likelihood of costly post-launch security fixes. Product managers should stay informed about emerging cybersecurity threats, compliance requirements, and industry best practices.

By including security features as core components of product development, they help create resilient products that earn and maintain user trust. They should also collaborate with compliance teams to ensure products meet relevant legal and industry standards such as GDPR, CCPA, PCI-DSS, and ISO 27001.

Risk Management and Threat Mitigation:

Product managers must actively participate in risk management processes, including threat modeling and vulnerability assessments. This involves identifying potential risks, evaluating their impact, and working with development teams to implement mitigation strategies.

They should also define security key performance indicators (KPIs) such as incident response times, number of vulnerabilities resolved, and product uptime. Monitoring these metrics helps ensure that security remains a continuous priority throughout the product lifecycle.

Customer Education and Transparency:

A crucial yet often overlooked responsibility is ensuring that customers understand the product's security features. Product managers should collaborate with marketing and customer support teams to create clear documentation, FAQs, and tutorials. Transparency about security practices builds trust and helps users adopt security-enhancing behaviors.

Case Study: Payment Gateway Security Upgrade

Imagine a leading payment gateway provider facing a significant surge in phishing attacks, posing a serious threat to its reputation and the trust its customers have in the platform. In such a scenario, the company's product management team takes on a crucial role in addressing the growing security concerns and leading a comprehensive overhaul of its security measures.

The company has been struggling with increasing instances of unauthorized access and fraud, which have resulted in a decline in customer confidence. With the pressure mounting to restore trust and safeguard sensitive user data, the product management team sets out to improve the platform's security while maintaining an intuitive user experience that would prevent any disruption in customer satisfaction.

To achieve this, the product management team implements a series of innovative and robust security measures. First, they introduce multi-factor authentication (MFA), which adds an additional layer of protection. By requiring users to verify their identities through multiple methods, the company ensures that even if one authentication method is compromised, unauthorized access can still be prevented.

Alongside MFA, the team integrates risk-based authentication into the platform. This dynamic system adjusts security protocols based on user behavior and transaction history, effectively adding a level of intelligence to the authentication process. By assessing the risk associated with each transaction, the system can implement stricter security measures for transactions that are deemed suspicious, while keeping the process seamless for legitimate users. This approach allows the platform to be both secure and user-friendly, reducing friction for everyday transactions while mitigating potential fraud risks.

To further enhance security, real-time fraud detection systems are implemented. These systems continuously monitor transaction patterns, immediately flagging any unusual activity for further investigation. By catching potential fraud in real time, the company can take swift action to block suspicious transactions before they are completed, ensuring a higher level of protection for users.

Recognizing the importance of maintaining a positive user experience, the product managers collaborate closely with UX designers to simplify the authentication process without compromising security. Their goal is to ensure that while the platform becomes more secure, users do not face unnecessary barriers when logging in or making transactions. By streamlining the authentication experience, the team ensures that security upgrades don't frustrate users or lead to a drop in engagement.

Additionally, the product management team launches a comprehensive educational campaign to keep users informed about the changes. Through tutorials, FAQs, and notifications, users are educated on how to securely manage their accounts and make the most of the newly implemented security features. This proactive approach reassures users that their safety is a priority and helps them feel more confident navigating the updated system.

After the security measures are implemented, post-implementation surveys reveal a noticeable increase in customer satisfaction. Many users express greater confidence in the platform's security, praising how the new features have made them feel more protected without complicating their experience.

Ultimately, the product management team's comprehensive approach not only mitigates the immediate security risks but also strengthens the company's market position. By improving the platform's security while maintaining a user-friendly experience, the team reinforces the brand's reputation as a trusted payment gateway provider, securing its place in a competitive and security-sensitive industry.

📖 Aligning Security with Core Product Goals

In today's competitive digital landscape, product managers must balance security features with the broader product goals of market expansion, customer engagement, and regulatory compliance. Security is not just a technical necessity but a strategic enabler that can drive customer trust and competitive advantage.

The Importance of Security-Product Alignment

Security features can no longer be treated as afterthoughts or isolated technical components. Instead, they should seamlessly integrate into the product's core functionalities. Poorly implemented security measures can hinder user experience, delay product launches, and increase operational costs. Conversely, well-aligned security features can enhance customer satisfaction, facilitate smoother market entries, and ensure long-term product viability.

Strategies for Effective Alignment
1. **Customer-Centric Design:**
    - **Conduct Usability Testing:** Security features should be intuitive and minimally intrusive. Product managers should collaborate with UX designers[172] to conduct usability tests, ensuring that authentication processes

---

[172]https://www.webstacks.com/blog/fintech-ux-design

like multi-factor authentication (MFA) or biometric verification are straightforward and user-friendly.

o **Feedback Loops:** Establish continuous feedback mechanisms to gather user input on security features, allowing for iterative improvements.

2. **Compliance-Driven Development:**

o **Regulatory Awareness:** Staying updated on industry-specific regulations such as GDPR, CCPA, and PCI-DSS is crucial. Product managers should work closely with legal and compliance teams to ensure all features meet the latest regulatory requirements.

o **Audit and Certification:** Pursue relevant certifications to build trust and demonstrate the product's commitment to security and compliance.

3. **Innovation Without Compromise:**

o **Leverage Advanced Technologies:** Invest in cutting-edge technologies like AI-powered fraud detection, machine learning for anomaly detection, and blockchain for secure transactions.

o **Proactive Threat Mitigation:** Use predictive analytics and real-time monitoring tools to identify and respond to potential threats before they escalate.

Case Study: Mobile Wallet Expansion

When a mobile wallet platform planned to expand into international markets, it faced complex challenges related to varying data protection laws and regional compliance standards. To navigate this, product managers initiated a multi-disciplinary collaboration involving legal, compliance, and engineering teams[173].

Key Actions Taken:

- **Regulatory Assessment:** Conducted a comprehensive review of regional data protection regulations, including local data residency and consent requirements.

- **Feature Customization:** Developed tailored security features such as localized encryption protocols and consent management systems.

- **Stakeholder Collaboration:** Organized regular cross-functional meetings to align product timelines with security and compliance milestones.

Outcome:

The integration of region-specific security features enabled the mobile wallet platform to meet compliance standards while launching on schedule. The result was enhanced user trust, a broader customer base, and a strengthened market presence.

By aligning security with core product goals, product managers can transform security from a regulatory hurdle into a strategic asset. A well-integrated approach ensures product growth, regulatory compliance, and customer loyalty, setting the stage for sustainable business success.

---

[173]

https://www.researchgate.net/publication/335953715_The_DAta_Protection_REgulation_COmpliance_Model

📖 Communicating Security Needs to Stakeholders

Communicating security priorities effectively is a critical responsibility for product managers, especially in today's evolving digital landscape. A well-informed and security-conscious leadership team can make timely decisions that significantly mitigate potential risks. Securing executive buy-in and fostering team collaboration hinges on clear, compelling, and actionable communication strategies.

Effective Communication Techniques

Data-Driven Presentations:

- Presenting relevant metrics is one of the most compelling ways to demonstrate the importance of security initiatives. Use data from past breaches, industry reports, or results from simulated attack scenarios to highlight potential vulnerabilities and the severity of their impact. For example, showing the financial loss caused by similar breaches in competitors' organizations can illustrate the potential consequences of inaction.

- Additionally, use key performance indicators (KPIs) like the number of blocked phishing attempts, average response time to incidents, or reduction in system downtimes after implementing security upgrades. Data visualizations, such as charts and infographics, can help simplify complex technical data for non-technical stakeholders.

Storytelling Approach:

- Narrative-driven communication can transform abstract security risks into relatable scenarios. Share real-world incidents where organizations faced severe consequences due to overlooked security issues. Personalize the story by framing what could happen if similar incidents occurred within the company.

- Consider beginning presentations with a hypothetical "day in the life" scenario of a security breach, describing the operational disruption, customer trust erosion, and financial losses that could occur. This approach fosters emotional engagement and urgency.

Stakeholder Workshops:

- Conduct interactive sessions to explain technical vulnerabilities in business-friendly terms. These workshops can include tabletop exercises simulating security incidents and discussing appropriate response strategies.

- Engage stakeholders by breaking them into cross-functional teams tasked with solving specific security challenges. This not only enhances understanding but also promotes a shared sense of responsibility toward security goals.

Case Study: Securing Board Approval for Security Funding

A product manager at a digital lending startup faced challenges securing a budget for improved cybersecurity infrastructure. The team decided to simulate the business impact of a potential ransomware attack, including data encryption disruptions, service outages, and regulatory fines.

The presentation began with a detailed timeline of a hypothetical attack scenario, showing how quickly the company could lose access to customer data and the likely financial repercussions.

Metrics such as projected revenue loss, potential customer churn, and legal liabilities were clearly outlined.

The product manager also highlighted industry-specific security incidents, referencing companies that faced public backlash and operational paralysis due to inadequate security measures. This comparison drove home the real-world relevance of the proposed budget.

As a result, the board promptly approved the requested funds for enhanced data encryption, real-time threat monitoring systems, and advanced incident response tools. The success of this approach underscored the power of blending data-driven insights, compelling storytelling, and engaging stakeholder interactions when communicating security needs[174].

📖 Balancing Security with Usability

Balancing security with usability[175] is a critical challenge in product development, especially in sectors where sensitive data is involved, such as financial technology (FinTech), healthcare, and e-commerce. Striking the right balance ensures that users feel safe without being hindered by cumbersome security measures. Product managers must adopt a dual-focus approach that prioritizes both aspects, recognizing that excessive security can drive users away while weak security exposes them to potential risks.

---

[174]https://userback.io/blog/cybersecurity-product-managers/
[175]https://medium.com/%40lerpalsocial/fintech-balancing-usability-and-security-12285e8df228

📖 Best Practices for Achieving Balance

User-Centered Security Design

- Incorporate frictionless authentication methods such as biometric verification, single sign-on (SSO), and passwordless logins.

- Conduct extensive A/B testing to determine which security features offer the best trade-off between protection and ease of use.

- Implement contextual authentication, where additional verification is only triggered during suspicious activities.

Progressive Security Measures

- Offer customizable security settings, allowing users to enable advanced features like two-factor authentication (2FA) or hardware security keys.

- Provide adaptive security that adjusts based on the user's behavior, device recognition, and location.

- Ensure that default security settings provide robust protection while keeping advanced features optional for tech-savvy or high-risk users.

Feedback Loops and Continuous Improvement

- Establish clear channels for gathering user feedback through surveys, in-app prompts, and customer support interactions.

- Monitor security-related drop-off rates and analyze user behavior data to identify friction points.

- Regularly update security features based on user input and emerging industry best practices.

Case Study: Streamlining Onboarding for a FinTech App

Let's assume a leading FinTech app is facing a significant challenge with high user drop-off rates during its onboarding process, primarily due to a time-consuming document verification procedure. Users are required to upload several identification documents, leading to frustration and abandonment.

To address this, the product manager conducts user interviews and analyzes data, discovering that the manual verification process is the primary bottleneck. In response, they decide to integrate third-party identity verification APIs that utilize AI-powered document scanning and facial recognition technology. This update significantly reduces onboarding time while enhancing fraud detection accuracy.

Furthermore, the app introduces a tiered onboarding process. New users can access basic features with minimal verification, while additional checks are required to unlock full functionality. This progressive approach strikes a balance between maintaining strong security measures and ensuring a smooth user experience, ultimately boosting customer retention and satisfaction.

By combining thoughtful design, adaptive security measures, and continuous user feedback, product managers can effectively create a seamless yet secure onboarding process that meets both security standards and user expectations.

📖 Advocating for Security Resources

Securing budget and technical resources for cybersecurity initiatives can be a challenging endeavor, especially in organizations where competing priorities exist. Product managers play a crucial role in this process by advocating for security investments through a value-driven approach that resonates with business stakeholders. Successfully securing funds requires blending technical expertise with business acumen to make a compelling case for cybersecurity enhancements.

Advocacy Tactics

To persuade executive leadership and other stakeholders, product managers should employ several key advocacy tactics:

ROI Calculations:
Demonstrating the potential return on investment (ROI) from cybersecurity expenditures can be highly effective. This involves presenting detailed cost-benefit analyses that compare the expenses of implementing security measures with the potential financial losses from breaches. Highlight how preventive measures can save the organization from costly incidents like data breaches, ransomware attacks, and operational downtime.

Competitive Analysis:
Understanding the competitive landscape can strengthen the business case for cybersecurity investments. Product managers should research competitors' security features and emphasize how maintaining robust cybersecurity can be a market differentiator. Showcasing industry benchmarks and best practices can underscore the importance of staying ahead in security capabilities.

Regulatory Compliance Costs:
Non-compliance with industry regulations and data privacy laws can result in significant financial penalties and reputational damage. Product managers should clearly outline the potential costs of failing to meet compliance requirements, emphasizing how proactive

investment in cybersecurity can mitigate these risks and ensure ongoing regulatory adherence.

Case Study: Securing Funds for a Security Overhaul

Consider the case of a cryptocurrency exchange that experienced a data breach, drawing regulatory scrutiny and eroding customer trust. The company's product manager recognized the urgent need for enhanced security measures and took a systematic approach to secure the required budget.

First, the product manager conducted a thorough risk assessment, highlighting critical vulnerabilities that led to the breach. They proposed investing in next-generation firewalls, advanced intrusion detection systems, and biometric login solutions to strengthen the platform's security infrastructure.

To reinforce the proposal, they presented a comprehensive ROI analysis, showing how the suggested upgrades could prevent future breaches, saving the company millions in potential losses from fraud, fines, and lost business. Additionally, they conducted a competitive analysis, demonstrating how leading competitors had already implemented similar security features, positioning themselves as trusted industry players.

The final step involved mapping the proposed security investments to regulatory compliance requirements. The product manager emphasized that failure to upgrade the company's cybersecurity infrastructure could result in steep fines and possible business restrictions from regulatory bodies.

The well-structured proposal convinced the executive team of the urgency and long-term value of the investment. As a result, the company secured a $2 million budget increase, enabling a comprehensive security overhaul that restored its reputation and fortified its defenses against future threats.

By applying these advocacy tactics, product managers can effectively navigate the complexities of securing cybersecurity funding, ensuring that their organizations remain resilient in an ever-evolving digital landscape.

📖 Integrating Security into Product Roadmaps

Embedding security within the product roadmap is essential for fostering a proactive rather than reactive approach to product development. By making security a fundamental component of the development process, companies can address potential vulnerabilities before they escalate into significant issues, ensuring long-term product stability and user trust.

Why Integrate Security into Product Roadmaps?

Security threats continue to evolve, making it crucial for organizations to anticipate and mitigate risks early. Reactive fixes can be costly, both financially and reputationally. Integrating security into the product roadmap ensures that security upgrades and best practices are considered from the initial planning phases, resulting in more robust and resilient products[176].

---

[176]https://safestack.io/blog/resources/product-management-cyber-security

Key Steps to Integration

Define Security Milestones:

Establish clear and achievable timelines for security updates and improvements. These milestones should align with product releases and be treated with the same priority as feature rollouts.

Cross-Functional Collaboration:

Involve security teams during the early stages of product development. Collaborative planning sessions between development, product management, and security teams help identify potential risks and design flaws before they become embedded in the product.

Risk Assessments and Audits:

Conduct regular security audits and risk assessments to identify vulnerabilities. Use these findings to adjust the product roadmap accordingly, prioritizing high-risk issues and integrating long-term security strategies.

Security by Design Principles:

Apply security best practices such as encryption, access control, and data masking during the design phase. This ensures that products are secure by default, reducing the need for extensive fixes later.

Continuous Monitoring and Feedback:

Implement monitoring tools to track security metrics and product performance. Use insights from real-world usage and incident reports to refine the product roadmap.

Compliance and Regulatory Alignment:

Ensure that security initiatives align with relevant industry regulations and compliance standards. Incorporating regulatory milestones within the roadmap can help avoid legal and financial penalties.

Case Study: Roadmap-Driven Security Evolution

A global remittance service faced growing cybersecurity threats due to the sensitive nature of financial transactions. To combat these challenges, the company implemented quarterly security reviews as part of its product roadmap. Each review assessed system vulnerabilities, compliance requirements, and potential new threats.

By proactively addressing issues uncovered during these reviews, the company consistently updated its encryption protocols, strengthened fraud detection mechanisms, and enhanced access controls. This roadmap-driven approach enabled the organization to maintain a competitive edge by staying ahead of emerging security challenges while boosting customer trust through transparent security practices.

Integrating security into product roadmaps is not just a best practice but a necessity in today's dynamic digital landscape. By embedding security milestones, fostering cross-functional collaboration, and conducting regular risk assessments, organizations can build secure, resilient products that meet user expectations and regulatory demands. A well-implemented security roadmap ultimately leads to reduced downtime, enhanced customer satisfaction, and sustained market competitiveness[177].

📖 Conclusion

Balancing security and usability is an evolving process that requires product managers to be strategic, data-driven, and collaborative. By embedding security into product development, aligning it with

---

[177]https://www.scienceopen.com/hosted-document?doi=10.14236%2Fewic%2FHCI2021.12

business goals, and advocating effectively, they can create products that are both secure and user-friendly. This dual focus not only safeguards the organization but also builds lasting trust with customers and stakeholders.

What We Learnt:

- **Product Managers' Role in Cybersecurity:** Product managers must integrate security into every stage of product development, balancing technical execution with business goals.

- **Prioritizing Security Features:** Security should enhance usability, not hinder it. Product managers must collaborate with developers and UX designers to implement features like biometric authentication and data encryption seamlessly.

- **Stakeholder Alignment:** Product managers act as bridges between technical teams, executives, marketing, and customers. Clear communication about security risks and priorities ensures organizational alignment.

- **Security-First Mindset:** Embedding security into the product roadmap from the start reduces future costs and builds user trust while ensuring compliance with industry standards.

- **Risk Management and Threat Mitigation:** Product managers should participate in risk management, conduct threat assessments, and monitor security KPIs like incident response time and resolved vulnerabilities.

- **Customer Education and Transparency:** Clear communication about security features builds customer trust and encourages secure product usage through tutorials, FAQs, and campaigns.

- **Aligning Security with Core Product Goals:** Security should support business expansion, customer engagement, and regulatory compliance while enhancing the product's overall competitiveness.

- **Communicating Security Needs:** Use data-driven presentations, storytelling, and stakeholder workshops to secure funding and resources for security initiatives.

- **Balancing Security with Usability:** Product managers must design intuitive security features while minimizing user friction through adaptive authentication, contextual security, and feedback-driven improvements.

- **Advocating for Security Resources:** Present ROI calculations, competitive analyses, and compliance risks to justify cybersecurity investments.

- **Integrating Security into Product Roadmaps:** Embedding security milestones, conducting regular audits, and fostering cross-functional collaboration ensure long-term product resilience and customer satisfaction.

CHAPTER 12
INCIDENT RESPONSE AND CRISIS MANAGEMENT

📖 Developing an Incident Response Plan

In the dynamic world of FinTech, where every second counts, having a well-defined Incident Response Plan (IRP) is not just recommended—it's essential. A robust IRP serves as a roadmap for identifying, managing, and mitigating security incidents, helping organizations minimize damage and recover swiftly. Without a clear incident response framework, companies risk extended downtime, financial losses, and reputational harm.

📖 Key Elements of an Incident Response Plan

Preparation:
Preparation is the foundation of an effective incident response plan (IRP)[178]. Organizations must proactively establish protocols, tools, and training mechanisms to ensure readiness when a cybersecurity incident occurs.

- **Establishing an Incident Response Team (IRT):** Organizations should create a dedicated IRT composed of IT, security, legal, public relations, and management

---

[178]https://www.deepthreatanalytics.com/network-security/effective-cybersecurity-incident-communication-plans/

personnel. Each member should have clearly defined roles and responsibilities to avoid confusion during an incident.

- **Conducting Regular Training Sessions:** Ongoing training ensures that team members remain current on the latest cybersecurity threats, technologies, and response procedures. Simulation exercises and tabletop exercises can improve readiness by testing various threat scenarios.

- **Creating a Detailed Incident Response Playbook:** A well-structured playbook should outline step-by-step actions for responding to different types of cybersecurity incidents, such as ransomware attacks, Distributed Denial of Service (DDoS) attacks, insider threats, and phishing attempts. The playbook should include escalation paths, communication protocols, and key contact information.

Identification:

Accurate and timely identification of security incidents is critical for minimizing potential damage.

- **Implementing 24/7 Monitoring Tools:** Organizations should deploy advanced security monitoring solutions, including intrusion detection systems (IDS), firewalls, and Security Information and Event Management (SIEM) platforms, to continuously monitor network traffic and system logs for suspicious activity.

- **Utilizing Threat Intelligence Feeds:** Real-time threat intelligence feeds provide up-to-date information on emerging threats, enabling faster detection and proactive defense measures.

- **Conducting Anomaly Detection:** Machine learning models and behavioral analytics tools can identify unusual patterns or deviations from normal operations, helping to distinguish between legitimate and potentially malicious activities.

Containment:

Containing the incident swiftly helps reduce its spread and minimizes the overall impact on the organization.

- **Initiating Containment Measures:** Immediate actions such as isolating affected systems, disconnecting compromised devices from the network, and blocking malicious IP addresses can help limit the incident's scope.

- **Enforcing Network Segmentation:** Network segmentation involves dividing the network into isolated segments, restricting the attacker's lateral movement, and limiting access to sensitive data and systems.

- **Applying Zero-Trust Principles:** A zero-trust security model ensures that every access request is verified and that access privileges are reduced during an active breach to prevent unauthorized access.

Eradication:

Once the threat is contained, eradicating the root cause ensures that similar incidents do not recur.

- **Removing Malicious Software:** Security teams should identify and eliminate all traces of malware, patch vulnerabilities, and reset or disable compromised accounts.

- **Conducting Malware Analysis:** Understanding the attack's technical details can help prevent future incidents. Comprehensive malware analysis should be performed to identify the infection vector and its full capabilities.

- **Engaging Third-Party Experts:** In complex incidents, third-party cybersecurity firms can conduct forensic investigations, providing deeper insights into the attack and ensuring thorough eradication.

Recovery:

Restoring normal business operations while ensuring data integrity and system reliability is crucial during the recovery phase.

- **Restoring Systems from Backups:** Organizations should use secure and verified backups to restore critical systems and data, ensuring that recovery efforts do not reinstate compromised files.

- **Performing Post-Recovery Testing:** Extensive testing should be conducted to validate system functionality and security. This includes penetration testing and vulnerability assessments.

- **Monitoring for Residual Threats:** Even after recovery, continuous monitoring is essential to detect any signs of residual threats or reinfection attempts.

Post-Incident Analysis:

Learning from the incident is key to improving future response efforts.

- **Hosting Post-Mortem Reviews:** A comprehensive review involving all stakeholders should be conducted to analyze the incident's timeline, actions taken, and the effectiveness of the response.

- **Documenting Lessons Learned:** Detailed reports should document findings, highlighting areas that need improvement and specifying new measures to prevent similar incidents.

- **Updating the IRP:** Based on the lessons learned, organizations should revise and update the incident response plan and provide refresher training to relevant personnel.

Case Study: Nvidia Ransomware Attack

In 2022, Nvidia faced a severe ransomware attack that encrypted sensitive customer data and disrupted its online services. The attack demonstrated the effectiveness of having a pre-established incident response plan[179].

- **Detection:** Within an hour of the breach, Nvidia's SIEM platform detected unusual activity, triggering an alert to the security team.

- **Containment:** The team quickly isolated the compromised servers, preventing the ransomware from spreading further into the network.

- **Eradication:** Security specialists identified the ransomware variant and removed it from affected systems. They patched vulnerabilities and strengthened endpoint security.

- **Recovery:** Thanks to secure offline backups, customer data and essential services were restored within 48 hours. Extensive testing confirmed the full functionality and security of all critical systems.

- **Post-Incident Analysis:** A thorough post-mortem review highlighted gaps in the bank's network segmentation and endpoint protection. The company enhanced its security

---

[179] https://www.cnbctv18.com/technology/nvidia-confirms-ransomware-attack-and-leak-of-data-hacking-the-hacker-didnt-help-12675082.htm

infrastructure, added new security monitoring tools, and conducted employee cybersecurity awareness training.

By following these key elements, organizations can create a robust incident response plan that minimizes business disruption, protects sensitive data, and strengthens overall cybersecurity resilience.

📖 Steps for Managing and Communicating Security Breaches

Effective incident management extends beyond technical containment; clear and transparent communication is equally critical. Organizations must be prepared to inform stakeholders, regulators, and customers promptly while balancing legal, regulatory, and reputational risks. A well-defined strategy ensures that breaches are managed efficiently while minimizing potential damages.

Incident Management Workflow

1. Detection and Reporting:

The first step in managing a security breach is swift detection and accurate reporting. Timely identification can significantly reduce potential damage. Organizations should implement the following practices:

- **Employee Training:** Ensure all employees can recognize and report suspicious activities quickly through a streamlined reporting process. This includes identifying phishing emails, unusual login attempts, and abnormal data access patterns.

- **Automated Alerts:** Deploy an automated alerting system that triggers immediate notifications to the Incident Response Team (IRT). These alerts should include critical information such as the type of threat detected, affected systems, and recommended initial response actions.

- **Reporting Channels:** Establish clear and accessible reporting channels, such as hotlines, internal portals, or mobile apps, to facilitate swift communication of potential breaches.

2. Incident Classification:

Proper classification of a security incident helps determine the level of response required. This process involves evaluating the incident's severity based on predefined criteria, including:

- **System Impact:** Assess the extent of system downtime or performance degradation.

- **Data Sensitivity:** Evaluate the type and sensitivity of the compromised data, such as personally identifiable information (PII), financial records, or intellectual property.

- **Breach Scope:** Determine the number of affected users, systems, or geographical locations.

- **Severity Levels:** Assign severity levels, ranging from low to critical, to guide response actions and resource allocation.

3. Internal Escalation:

Once the incident is classified, organizations must escalate the issue to relevant internal teams to coordinate an effective response. Key steps include:

- **Team Notifications:** Notify IT, legal, compliance, public relations, and executive leadership immediately.

- **Crisis Management Activation:** For high-severity incidents, activate the crisis management team and appoint an incident manager to lead response efforts.

- **Cross-Department Collaboration:** Encourage close collaboration between technical and non-technical teams to ensure a unified response.

4. Stakeholder Communication:

Transparent communication with affected stakeholders builds trust and mitigates potential fallout. Organizations should follow these best practices:

- **Customer Notifications:** Inform affected customers promptly, outlining what happened, the data involved, and immediate protective steps they should take.

- **Regulatory Compliance:** Notify relevant regulators within legally mandated timeframes, such as 72 hours under GDPR.

- **Business Partner Alerts:** Communicate with business partners and vendors to prevent the breach from spreading through interconnected systems.

- **Secure Updates:** Provide updates through secure and trusted communication channels to maintain confidentiality.

5. Media and Public Relations:

Public perception can be significantly impacted by how an organization handles breach-related communication. Consider the following:

- **Spokesperson Designation:** Appoint a trained spokesperson authorized to communicate with the media.

- **Fact-Based Messaging:** Craft clear, fact-based public statements, avoiding speculative responses.

- **Message Consistency:** Ensure consistent messaging across all communication channels, including press releases, social media, and company websites.

- **Reputation Management:** Collaborate with PR agencies, if necessary, to manage public sentiment and reduce reputational damage.

6. Documentation and Compliance:

Thorough documentation and compliance practices ensure accountability and facilitate post-incident evaluations. Essential practices include:

- **Incident Records:** Maintain comprehensive incident records, including timelines, actions taken, and communications.

- **Compliance Checks:** Ensure adherence to data protection laws such as GDPR, CCPA, PCI-DSS, and other industry-specific regulations.

- **Post-Incident Review:** Conduct post-mortem analyses to identify gaps in response efforts and implement improvements.

Case Study: Global Payments Data Breach

In 2021, Global Payments experienced a significant data breach affecting millions of customer records. Due to a lack of coordinated communication, the company faced intense public backlash. Customers were left uninformed for days, causing widespread dissatisfaction and reputational damage[180].

---

[180] https://studymoose.com/case-analysis-global-payments-breach-essay

Realizing the critical role of transparent communication, the company revamped its incident management and communication protocols. It established a dedicated incident response team, improved real-time communication channels, and mandated regular public updates during crisis situations. As a result, subsequent incidents were handled with greater efficiency, restoring customer trust and mitigating further damage.

By learning from such real-world examples, organizations can strengthen their incident management strategies, ensuring both effective containment and transparent communication when managing security breaches[181].

📖 Learning from Past Breaches: Turning Incidents into Improved Security Practices

Every cybersecurity breach presents a critical learning opportunity. While breaches[182] can be disruptive and costly, they also offer invaluable lessons that can drive significant improvements in an organization's security posture. By analyzing real-world incidents, organizations can identify vulnerabilities, enhance processes, and strengthen their overall security culture. Transforming breaches into actionable insights requires a proactive approach grounded in systematic reviews, updated policies, continuous training, and collaboration with industry partners.

---

[181]https://www.securityweek.com/global-payments-data-breach-cost-nearly-85-million/
[182]https://www.bpm.com/insights/cyber-incident-response/

How to Turn Incidents into Actionable Insights?

1. Conduct Post-Incident Reviews

A thorough post-incident review should be a cornerstone of every incident response process. These reviews provide clarity on what happened, how the breach was handled, and what can be improved.

- **Host Detailed Debriefs:** Bring together all members of the incident response team for a comprehensive debrief. Each team member should share their perspective on how the breach unfolded and how it was managed.

- **Identify Successes and Gaps:** Highlight actions that were effective, as well as areas that need improvement. Document lessons learned and assign action items with deadlines to prevent recurrence.

- **Document Findings:** Create a detailed report outlining the breach timeline, response actions, and recovery process. This report should be shared with relevant stakeholders to ensure transparency and facilitate organizational learning.

2. Identify Root Causes

Understanding the root cause of a breach is essential for addressing underlying issues that may lead to future incidents.

- **Pinpoint Technical Flaws:** Identify vulnerabilities in system configurations, outdated software, or security misconfigurations.

- **Recognize Human Errors:** Evaluate whether employee mistakes, such as clicking on phishing links or using weak passwords, contributed to the breach.

- **Address Procedural Gaps:** Review security processes and protocols for weaknesses, such as inadequate incident response procedures or insufficient monitoring.

- **Utilize Forensic Analysis:** Engage third-party experts if necessary to conduct forensic investigations, ensuring an unbiased analysis of the breach.

3. Update Security Policies

Effective security policies serve as the foundation of a strong cybersecurity program. Breaches often reveal policy shortcomings that require immediate updates.

- **Revise Access Controls:** Ensure that employees only have access to systems and data relevant to their job roles. Implement role-based access and least privilege principles.

- **Strengthen Authentication Protocols:** Enhance authentication procedures by enforcing multi-factor authentication (MFA), complex password requirements, and regular password changes.

- **Improve Data Protection Standards:** Update data encryption policies to safeguard sensitive information, both in transit and at rest. Regularly audit encryption protocols for compliance with current standards.

4. Invest in Continuous Training

A well-informed workforce is a crucial line of defense against cyber threats. Training should be ongoing and adaptive to the evolving threat landscape.

- **Conduct Regular Training Sessions:** Offer cybersecurity training tailored to employees' roles, emphasizing best practices and emerging threats.

- **Simulate Phishing Attacks:** Launch simulated phishing campaigns to raise employee awareness and reinforce the importance of vigilance.

- **Promote Security Culture:** Encourage a culture of security by recognizing employees who excel in following cybersecurity best practices.

5. Simulate Breaches Regularly

Testing incident response plans through simulations helps ensure that teams are ready to handle real-world threats.

- **Conduct Tabletop Exercises:** Simulate breach scenarios in a controlled environment to test response strategies and identify areas for improvement.

- **Perform Penetration Testing:** Hire ethical hackers to conduct penetration tests and expose system vulnerabilities before attackers can exploit them.

- **Develop Red Team vs. Blue Team Exercises:** Engage in red team vs. blue team simulations to evaluate defenses under realistic attack conditions, fostering hands-on learning and collaboration.

6. Leverage Threat Intelligence

Staying informed about emerging threats allows organizations to anticipate potential attacks and adjust their security measures accordingly.

- **Subscribe to Threat Intelligence Feeds:** Use industry-specific threat intelligence services to receive updates on current threats, vulnerabilities, and attack methods.

- **Collaborate with Industry Partners:** Join cybersecurity information-sharing communities, collaborate with other organizations, and maintain relationships with law enforcement for real-time threat sharing.

Case Study: FinTechCorp's Insider Threat Mitigation

FinTechCorp faced a serious insider threat when a former employee attempted unauthorized access to sensitive systems after termination. This incident prompted a comprehensive review and major security enhancements[183].

- **Strengthened Offboarding Processes:** The company revised its offboarding procedures, ensuring immediate revocation of all system access upon employee departure.

- **Automated Access Revocation:** FinTechCorp automated the access revocation process through identity and access management (IAM) solutions, reducing reliance on manual interventions.

- **Enhanced Privileged Account Monitoring:** Stricter monitoring of privileged accounts was implemented using real-time monitoring tools and anomaly detection systems.

By embedding these lessons into their cybersecurity strategy, FinTechCorp reduced incident response times, minimized potential damages, and built a more resilient security posture. This proactive approach can help organizations across industries better withstand future threats and ensure long-term operational security[184].

---

[183]https://finsecdevinsights.com/2024/01/incident-response-101-strategies-for-swift-action-in-fintech-security/
[184]https://www.neumetric.com/insider-threats-mitigation-strategies/

📖 Conclusion

Turning breaches into actionable insights is essential for modern cybersecurity resilience. Through post-incident reviews, root cause analysis, policy updates, training, breach simulations, and threat intelligence sharing, organizations can continuously improve their security defenses. While breaches may be inevitable, a well-prepared organization can limit their impact and emerge stronger, more secure, and ready for the challenges ahead[185].

💡 What We Learnt:

- **Having an Incident Response Plan (IRP):** Minimizes damage, ensures quick recovery, and protects organizational reputation.

- **Establishing an Incident Response Team (IRT):** Helps coordinate roles and responsibilities during security incidents.

- **Conducting Regular Training:** Keeps the team updated on cybersecurity threats and response procedures.

- **Creating a Detailed Incident Response Playbook:** Provides step-by-step guidance for handling various cybersecurity incidents.

- **Using Monitoring Tools and Threat Intelligence Feeds:** Enables quick detection of potential security threats.

---

[185]https://www.vaia.com/en-us/explanations/computer-science/fintech/incident-response-planning/

- **Isolating Affected Systems:** Prevents the spread of malicious activities during a breach.

- **Applying Zero-Trust Principles:** Enhances security by verifying every access request during an incident.

- **Removing Malicious Software:** Ensures complete elimination of malware and patches vulnerabilities.

- **Restoring Systems from Backups:** Helps resume business operations while preserving data integrity.

- **Conducting Post-Incident Reviews:** Identifies gaps in the response process and recommends improvements.

- **Classifying Security Incidents:** Guides response actions based on the severity and impact of the breach.

- **Notifying Relevant Teams:** Ensures internal teams collaborate effectively for incident resolution.

- **Informing Stakeholders Promptly:** Builds trust and mitigates potential reputational damage.

- **Appointing a Trained Spokesperson:** Ensures consistent, fact-based communication with the public.

- **Maintaining Incident Records:** Supports compliance and facilitates future incident investigations.

- **Updating Security Policies:** Strengthens access controls, authentication protocols, and data protection standards.

- **Conducting Continuous Training:** Keeps employees aware of evolving cybersecurity threats.

- **Simulating Breach Scenarios:** Tests incident response readiness and highlights areas for improvement.

- **Leveraging Threat Intelligence:** Anticipates potential attacks and enhances defensive measures.

CHAPTER 13
BUILDING A SECURITY-CONSCIOUS PRODUCT CULTURE

📖 Introduction

In the ever-evolving FinTech landscape, cybersecurity is not merely a technical concern but a cultural imperative that shapes the very foundation of successful product development. A security-conscious product culture ensures that security becomes an integral part of every development stage, from concept creation to post-launch maintenance. It involves embedding security principles into the company's DNA, fostering a collective responsibility among teams, and aligning business goals with cybersecurity priorities. This chapter explores comprehensive strategies for fostering a security-first mindset at every organizational level, ensuring that cybersecurity is treated not as a barrier but as an enabler of innovation and trust.

📖 Establishing a Security-Conscious Product Culture

1. Defining Core Security Values

- **Leadership Commitment:** Leadership must actively champion cybersecurity by allocating resources, setting policies, and participating in security-related initiatives.

Successful examples include Microsoft's "Security Development Lifecycle," where executive support led to enterprise-wide security adoption, setting an industry benchmark.

- **Security as a Shared Responsibility:** Every team member, regardless of role, should understand that they play a part in maintaining product security. This concept can be seen in companies like Google, where security-awareness programs are integrated into daily work routines, reinforcing a shared responsibility culture.

2. Security-Driven Development Practices

- **Security by Design:** Embed security principles into every phase of product development. For instance, consider Facebook's "Hacktober" initiative, where developers identify and fix vulnerabilities through gamified security challenges.

- **Threat Modeling:** Conduct regular threat assessments to anticipate and mitigate risks. Amazon's approach to threat modeling includes "Red Team" exercises, where internal teams simulate attacks to uncover weaknesses before malicious actors can exploit them.

- **Code Reviews and Audits:** Implement routine code reviews and third-party security audits. GitHub has made notable strides by incorporating automated code scanning tools like Dependabot, ensuring that vulnerabilities are detected and addressed swiftly.

3. Security Policies and Guidelines

- **Clear Documentation:** Establish and regularly update policies covering data protection, incident response, and compliance. Companies like IBM maintain extensive internal security policies accessible through secure portals, keeping employees informed and aligned.

- **Access Controls:** Implement strict access control policies, ensuring data is only accessible to authorized personnel. For example, Netflix uses a zero-trust security model that verifies each access request regardless of its origin, significantly reducing insider threats.

Case Study: Capital One Data Breach (2019)

In 2019, Capital One experienced a data breach impacting over 100 million customers due to a misconfigured firewall in its cloud infrastructure[186]. The incident underscored the importance of access control policies and proactive system audits. In response, the company revamped its cybersecurity strategy by enhancing data encryption protocols, strengthening its identity access management framework, and conducting continuous security training across teams[187].

By integrating such best practices and learning from real-world incidents, organizations can build a resilient security-conscious culture that proactively addresses evolving cyber threats[188].

---

[186]
https://www.researchgate.net/publication/340012934_A_Case_Study_of_the_Capital_On e_Data_Breach
[187] https://www.cnn.com/2019/07/29/business/capital-one-data-breach/index.html
[188] https://www.microsoft.com/en-us/securityengineering/sdl/threatmodeling

📖 Cross-Functional Training and Empowering Teams

To maintain a competitive edge in today's rapidly evolving business environment, organizations must adopt a cross-functional approach to training and empower their teams with relevant skills. This involves creating tailored training programs, establishing a Security Champions Program, and fostering collaborative security initiatives.

1. Training Programs

Role-Specific Training

Tailoring training sessions based on team roles ensures employees gain relevant and actionable knowledge. Key focus areas include:

- **Development Teams:** Provide secure coding workshops, version control management, and software development lifecycle (SDLC) best practices.

- **Marketing Teams:** Conduct sessions on data privacy regulations, social media security, and safe campaign execution.

- **Customer Service Teams:** Offer training on handling sensitive customer data, recognizing social engineering tactics, and following secure communication protocols.

Case Study: A global e-commerce company reduced customer service data breaches by 30% after introducing role-specific security training. Customer service agents learned to identify and mitigate phishing scams, leading to improved customer trust[189].

---

[189]https://ttms.com/cybersecurity-training-how-e-learning-can-help-to-keep-your-company-safe/

Cybersecurity Workshops

Conducting hands-on workshops equips employees with practical skills to handle cybersecurity threats. Suggested workshop topics include:

- **Phishing Prevention:** Simulated phishing campaigns followed by debrief sessions.

- **Password Management:** Training on creating and managing secure passwords using password managers.

- **Secure Coding Practices:** Coding competitions focused on writing secure and efficient code.

Case Study: A financial services firm successfully reduced incidents of credential theft by 40% after conducting monthly phishing awareness workshops[190].

2. Security Champions Program

Identify Security Advocates

Security champions are team members passionate about cybersecurity. Their responsibilities include:

- Promoting security awareness within their teams.

- Serving as the first point of contact for security-related queries.

- Collaborating with the IT and security teams to address vulnerabilities.

---

[190]https://aag-it.com/the-latest-phishing-statistics/

Continuous Learning

Encourage security champions to participate in industry conferences, pursue certifications like CISSP or CEH, and engage in security forums.

Case Study: A tech startup appointed security champions within its development and product teams. As a result, the organization experienced a 25% reduction in application vulnerabilities, thanks to proactive threat detection by trained staff[191].

3. Collaborative Security Initiatives

Security Hackathons

Hosting internal hackathons encourages creativity and problem-solving. Suggested formats include:

- **Bug Bounty Simulations:** Employees compete to find and fix bugs.

- **Innovation Challenges:** Teams build secure apps or design new security features.

Case Study: An IT firm discovered critical security flaws in its flagship product during an internal hackathon, allowing for timely resolution before the product launch[192].

Cross-Team Communication

Regular communication between development, IT, and security teams helps resolve security issues faster. Suggested practices include:

- **Weekly Security Roundtables:** Teams share updates on current threats and mitigation efforts.

---

[191]https://cvcompiler.com/cybersecurity-resume-examples
[192]https://www2.deloitte.com/content/dam/Deloitte/in/Documents/risk/in-risk-RegTech-Gaining-momentum-noexp.pdf

- **Incident Response Drills:** Cross-functional teams practice coordinated responses to simulated security breaches.

Case Study: A healthcare organization improved its incident response time by 45% after instituting bi-weekly security meetings and emergency response simulations[193].

By investing in cross-functional training and fostering collaboration, organizations can create a security-first culture that proactively mitigates risks while enhancing productivity. These initiatives build a more resilient and knowledgeable workforce, ready to tackle modern business challenges.

📖 Building Resilience for Future Challenges

Resilience in the face of evolving technological, environmental, and geopolitical challenges is essential for organizations aiming to secure long-term success. This involves proactive strategies in incident response, continuous monitoring, and adapting to emerging threats. Here's a comprehensive guide with detailed measures and relevant case studies.

1. Incident Response and Crisis Management

Incident Response Plans

- **Development and Documentation:** Organizations should create comprehensive incident response plans that outline roles, responsibilities, and step-by-step procedures. These plans should cover various scenarios such as data breaches, ransomware attacks, and natural disasters.

---

[193]https://www.phe.gov/preparedness/planning/cybertf/documents/report2017.pdf

- **Regular Updates:** Review and update response plans at least quarterly or after major incidents to reflect evolving threats.

Case Study: In 2021, a multinational financial services firm suffered a data breach due to a third-party vendor compromise. The firm's well-documented incident response plan enabled a swift response, including system isolation, data recovery, and regulatory notification, limiting damage and reducing recovery time[194].

Simulation Drills
- **Routine Exercises:** Conduct simulated crisis scenarios such as phishing attacks or system outages to evaluate organizational readiness.

- **Stakeholder Involvement:** Include cross-functional teams such as IT, legal, PR, and executive management.

Case Study: A global technology company runs annual red-team-blue-team simulations to assess cybersecurity defenses. During one exercise, the red team uncovered a major vulnerability in the company's cloud infrastructure, prompting immediate corrective action[195].

2. Continuous Monitoring and Risk Management

Real-Time Threat Monitoring
- **Advanced Tools:** Use AI-powered tools for real-time monitoring, anomaly detection, and automated alerts.

---

[194] https://www.gov.uk/government/statistics/cyber-security-breaches-survey-2024/cyber-security-breaches-survey-2024
[195]https://www.mitre.org/sites/default/files/2022-04/11-strategies-of-a-world-class-cybersecurity-operations-center.pdf

- **Threat Analysis:** Continuously assess threat data to identify patterns and potential breaches before they occur.

Case Study: A healthcare provider deployed an AI-based monitoring tool that detected unusual login patterns. Early intervention prevented a major ransomware attack that could have disrupted patient services[196].

Regular Audits and Compliance Checks

- **Scheduled Audits:** Conduct audits at regular intervals to assess policy adherence and uncover vulnerabilities.

- **Compliance Frameworks:** Align with industry-specific regulations such as GDPR, HIPAA, or PCI-DSS.

Case Study: An e-commerce giant faced regulatory penalties after neglecting compliance audits. Post-incident, the company implemented quarterly audits, reducing compliance breaches and restoring stakeholder trust[197].

3. Adapting to Emerging Threats

Threat Intelligence Sharing

- **Collaboration Networks:** Join industry cybersecurity alliances and threat intelligence platforms.

- **Information Sharing:** Share critical threat data with partners while adhering to privacy regulations.

Case Study: The banking sector has seen significant success through shared threat intelligence platforms. In one instance, early

---

[196]https://www.intechopen.com/online-first/1196803
[197]https://agp.gov.pk/SiteImage/Policy/CONSLIDATED%20AUDIT%20REPORT%20F EDERAL%20AY2023%2024.pdf.pdf

detection of a global malware outbreak through such a platform enabled multiple banks to strengthen their defenses preemptively.

Innovative Security Solutions

- **Cutting-Edge Technologies:** Invest in next-generation technologies such as blockchain for secure transactions, zero-trust architectures for identity management, and post-quantum cryptography to counter future quantum computing threats.

- **Research and Development:** Allocate resources to in-house research teams for developing proprietary security technologies.

Case Study: A defense contractor developed a zero-trust security model, integrating biometric authentication and blockchain-based identity verification. This approach significantly reduced insider threats and improved overall security[198].

Building resilience against future challenges requires a comprehensive strategy that combines incident response planning, continuous monitoring, and adaptive security innovations. By learning from real-world case studies and implementing best practices, organizations can navigate uncertainties while securing their digital and operational infrastructures[199].

📖 Conclusion

Building a security-conscious product culture is a journey, not a destination. It requires sustained commitment, cross-functional collaboration, and a proactive approach to evolving threats. By

---

[198]https://www.techscience.com/cmc/online/detail/22270/pdf
[199]https://safety.google/cybersecurity-advancements/cyberworkforce/

embedding security into the organization's DNA, FinTech companies can protect their products, users, and reputations while fostering a culture of trust and innovation.

What We Learnt:

- **Cybersecurity as a Cultural Imperative:** Security should be embedded in the company's DNA, aligning business goals with security priorities.

- **Leadership Commitment:** Executive support drives enterprise-wide adoption of security practices.

- **Security as a Shared Responsibility:** Every team member, regardless of role, must contribute to product security.

- **Security by Design:** Incorporate security principles into every stage of product development.

- **Threat Modeling:** Regular threat assessments help anticipate and mitigate risks.

- **Code Reviews and Audits:** Continuous reviews and third-party audits ensure timely vulnerability detection.

- **Clear Security Policies:** Establish and update data protection, access control, and incident response policies.

- **Cross-Functional Training:** Role-specific training ensures employees have relevant security skills.

- **Security Champions Program:** Identify and train advocates within teams to promote security awareness.

- **Collaborative Security Initiatives:** Encourage teamwork through hackathons, incident response drills, and regular security meetings.

- **Incident Response and Crisis Management:** Create, document, and update incident response plans for various security scenarios.

- **Simulation Drills:** Conduct routine crisis simulations involving cross-functional teams.

- **Real-Time Monitoring and Audits:** Use AI-powered monitoring tools and schedule regular audits to identify and fix vulnerabilities.

- **Compliance with Industry Standards:** Align security practices with regulatory frameworks like GDPR, HIPAA, and PCI-DSS.

- **Threat Intelligence Sharing:** Collaborate with industry partners to stay ahead of emerging threats.

- **Innovative Security Solutions:** Invest in advanced technologies like zero-trust architecture, blockchain, and post-quantum cryptography.

- **Continuous Adaptation:** Stay proactive by updating security strategies and learning from real-world incidents.

PART 6

EMERGING TRENDS AND BEST PRACTICES

CHAPTER 14
EMERGING TRENDS IN FINTECH CYBERSECURITY

The financial technology (FinTech) industry has experienced exponential growth in recent years, driven by technological innovations and a growing demand for digital financial services. However, this rapid expansion has made FinTech an attractive target for cybercriminals. In this chapter, we explore key emerging trends in FinTech cybersecurity, including quantum computing's impact on cryptography, the rise of AI-driven cyberattacks, and security challenges in decentralized finance (DeFi). We will also examine case studies that highlight real-world implications.

📖 Quantum Computing and Its Impact on Cryptography and FinTech Security

Quantum computing[200] represents a transformative leap in computational power, moving beyond the capabilities of classical computing systems. While classical computers operate using binary bits that represent either a 0 or a 1, quantum computers utilize quantum bits or qubits. Qubits can exist in multiple states simultaneously due to quantum properties such as superposition and entanglement. This unique capability allows quantum computers to

---

[200]https://wjaets.com/sites/default/files/WJAETS-2024-0333.pdf

perform complex calculations at speeds that are unattainable for even the most advanced classical supercomputers.

Implications for Cryptography

The emergence of quantum computing presents a significant challenge to the field of cryptography, which underpins the security infrastructure of FinTech systems. Many of today's encryption protocols rely on the computational difficulty of certain mathematical problems, such as factoring large prime numbers or computing discrete logarithms. Quantum computers, however, could disrupt this foundation by rendering these problems solvable within a feasible time frame.

Quantum Threat: Shor's Algorithm

One of the most concerning algorithms in the context of quantum computing is Shor's algorithm, which efficiently factors large integers. Modern cryptographic systems like RSA and Elliptic Curve Cryptography (ECC) depend on the practical impossibility of factoring such numbers using classical methods. A sufficiently powerful quantum computer running Shor's algorithm could break these encryption methods, exposing sensitive financial data, transactions, and personal information[201].

> 🔍 For example, RSA encryption, widely used in securing online banking, blockchain transactions, and digital signatures, could be rendered obsolete if quantum computers become capable of factoring 2048-bit keys in a matter of hours or even minutes.

---

[201] https://www.wsj.com/articles/companies-prepare-to-fight-quantum-hackers-c9fba1ae

Post-Quantum Cryptography (PQC)

To counteract the potential quantum threat, cryptographers have been working on post-quantum cryptography (PQC) — a new generation of encryption algorithms designed to be resistant to quantum attacks. These algorithms rely on mathematical problems that even quantum computers find challenging, such as lattice-based, hash-based, and multivariate polynomial cryptography.

The U.S. National Institute of Standards and Technology (NIST) has been leading efforts to standardize PQC algorithms. After an extensive evaluation process involving experts from around the globe, NIST announced its first set of PQC standards in 2022. These standards aim to replace vulnerable protocols like RSA and ECC with quantum-resistant alternatives, ensuring long-term data security[202].

📖 Case Study: Google's Quantum Supremacy Announcement

In 2019, Google claimed to have achieved quantum supremacy by performing a specific calculation in 200 seconds that would have taken the world's most advanced classical supercomputer approximately 10,000 years. Although the task was highly specialized and unrelated to cryptography, it demonstrated the disruptive potential of quantum computing[203].

This milestone served as a wake-up call for FinTech companies and governments alike, emphasizing the urgent need to adopt quantum-

---

[202] https://www.nist.gov/news-events/news/2024/08/nist-releases-first-3-finalized-post-quantum-encryption-standards
[203] https://arxiv.org/pdf/2210.12753

resistant cryptographic protocols. Financial institutions have since accelerated research into quantum-safe security measures, understanding that a sudden breakthrough in quantum computing could compromise today's encryption standards[204].

Industry Response and Preparation

Recognizing the existential risk posed by quantum computing, major financial institutions and technology firms have initiated collaborations aimed at developing quantum-safe security frameworks:

- **JPMorgan Chase:** This financial giant has partnered with quantum computing firms and academic institutions to explore quantum-safe cryptography, ensuring the protection of sensitive client data and secure transactions.

- **IBM and Microsoft:** Tech leaders like IBM and Microsoft have invested heavily in quantum research, focusing on developing hybrid cloud solutions with integrated quantum-safe protocols.

- **Blockchain Platforms:** Platforms like Ethereum have begun exploring quantum-resistant consensus mechanisms, aware that a quantum-capable adversary could disrupt blockchain networks by reversing transactions or forging digital signatures.

The Road Ahead: Balancing Innovation and Security

The race toward quantum computing supremacy is also a race for cybersecurity readiness. Financial institutions, governments, and tech companies must work together to ensure that when quantum

---

[204]https://link.springer.com/article/10.1007/s43681-024-00427-4

computing becomes mainstream, FinTech systems are prepared with robust, quantum-resistant cryptographic standards.

In the coming years, the transition to quantum-safe cryptography will become not just a competitive advantage but a security necessity. Standardization efforts, increased investment in quantum research, and proactive implementation of PQC algorithms will shape the future of FinTech security in a post-quantum world.

📖 The Rise of AI-Driven Cyberattacks and Counter-Strategies

Artificial Intelligence (AI) has revolutionized the cybersecurity landscape, serving as both a powerful shield and a formidable weapon. As cybercriminals adopt AI to launch sophisticated attacks, organizations must counteract with equally advanced AI-driven defenses. This escalating arms race has reshaped how the financial and tech sectors perceive and manage cybersecurity risks.

AI-Powered Threats

AI-driven cyberattacks involve the use of intelligent algorithms capable of adapting, learning, and bypassing traditional security systems. Common tactics include deepfakes, automated phishing, and adversarial machine learning attacks.

1. Deepfakes and Social Engineering
Deepfake technology has evolved into a significant cybersecurity threat. Using AI-powered synthetic media, attackers create convincing audio and video impersonations. For example, in 2019, a UK-based energy firm's CEO was duped into transferring

€220,000 to a fraudulent account after a deepfake audio file mimicked his boss's voice[205].

Such attacks enable large-scale financial fraud, stock market manipulation, and corporate espionage. In response, companies have begun investing in deepfake detection algorithms to verify the authenticity of communications[206].

2. Automated Phishing Attacks

Traditional phishing campaigns rely on human-created emails, which are easier to detect. AI-powered bots, however, can craft highly personalized phishing messages by scraping social media profiles and learning about a target's preferences, contacts, and recent activities.

In 2021, Microsoft reported that AI-powered phishing campaigns targeted over 30,000 companies worldwide. Attackers used natural language processing (NLP) models to generate contextually appropriate emails that bypassed conventional spam filters[207].

3. Adversarial Machine Learning

Attackers can manipulate AI systems by poisoning the training datasets used by fraud detection and anti-malware programs. In adversarial machine learning attacks, malicious data injected into a system's training set can cause the model to make incorrect decisions.

🔍 For example, in 2022, a global payment processor faced an adversarial attack where fraudulent transactions were classified as legitimate due to manipulated training data. This breach exposed the financial institution to millions of

---

[205]https://www.itjones.com/blogs/ai-deepfakes-and-the-evolution-of-ceo-fraud
[206]https://www.wa.gov.au/system/files/2024-10/case.study_.deepfakes.docx
[207] https://www.microsoft.com/en-us/microsoft-365-life-hacks/privacy-and-safety/how-ai-changing-phishing-scams

dollars in losses before the vulnerability was detected and corrected[208].

Defensive AI Solutions

To combat AI-driven threats, cybersecurity firms have developed advanced defensive AI technologies that enhance security infrastructure through real-time threat detection, automated responses, and continuous monitoring[209].

1. Behavioral Analytics

AI-based behavioral analytics systems monitor users' digital footprints to identify anomalies that could indicate fraudulent activity. Banks and FinTech firms deploy these systems to detect suspicious login attempts, irregular transactions, or unauthorized access.

Q For example, PayPal's AI-powered fraud detection platform analyzes billions of transactions in real-time, flagging anomalies based on historical user behavior. This proactive monitoring prevents potential breaches before they occur[210].

2. Threat Intelligence Platforms

Global threat intelligence platforms powered by AI aggregate and analyze cybersecurity data from thousands of sources, predicting emerging threats. They generate risk scores and deliver actionable insights to cybersecurity teams.

Companies like IBM and FireEye have integrated AI into their threat intelligence platforms, allowing organizations to anticipate and

---

[208]https://www.researchgate.net/publication/375062115_A_Comprehensive_Analysis_of_High-Impact_Cybersecurity_Incidents_Case_Studies_and_Implications
[209]https://link.springer.com/article/10.1007/s43681-024-00427-4
[210]https://www.paypal.com/us/brc/article/data-analytics-fraud-management

block potential attacks. This predictive approach has proven effective in reducing response times and mitigating breaches.

3. AI-Powered Security Operations Centers (SOCs)

Financial firms are increasingly deploying AI-enhanced SOCs to automate routine security tasks such as incident detection, response orchestration, and vulnerability management. AI-powered SOCs improve efficiency and reduce human workload.

JPMorgan Chase's AI-driven SOC analyzes thousands of alerts daily, filtering out false positives and allowing security analysts to focus on genuine threats. This approach minimizes downtime and enhances threat response times.

Case Study: The 2020 Twitter Bitcoin Scam

One of the most high-profile cyberattacks in recent years was the July 2020 Twitter Bitcoin scam. Cybercriminals compromised several high-profile Twitter accounts, including those of Elon Musk, Barack Obama, and Bill Gates, posting fraudulent messages promoting a Bitcoin giveaway[211].

While the attack primarily involved traditional social engineering, cybersecurity experts believe that AI-powered social engineering bots played a critical role in amplifying the attack's reach. These bots automatically retweeted and liked the fraudulent posts, boosting their visibility and creating a false sense of legitimacy.

The breach resulted in over $100,000 in fraudulent Bitcoin transfers and exposed Twitter's internal security flaws. In response, Twitter implemented stricter internal controls, including AI-driven

---

[211]https://thehackernews.com/2023/05/mastermind-behind-twitter-2020-hack.html

monitoring tools to detect suspicious account activity in real-time[212].

Industry Response

The cybersecurity industry has responded aggressively to the rising tide of AI-driven attacks. Leading cybersecurity firms like Darktrace, CrowdStrike, and Palo Alto Networks have developed AI-enhanced security products that detect, mitigate, and respond to cyber threats.

Major financial institutions such as HSBC and CitiBank have heavily invested in AI-powered fraud detection systems to secure their operations. These banks leverage AI to monitor billions of transactions, flagging and blocking potential fraud attempts within seconds.

Conclusion

The rise of AI-driven cyberattacks has transformed the cybersecurity landscape, necessitating a shift from reactive to proactive defense strategies. As cybercriminals become more sophisticated, organizations must stay ahead by adopting advanced AI-powered solutions. A comprehensive approach combining technological innovation, industry collaboration, and continuous monitoring will be essential to counter the ever-evolving threat landscape.

---

[212]https://www.dfs.ny.gov/Twitter_Report

📖 DeFi, Decentralized Finance, and Associated Security Risks

Decentralized Finance (DeFi) leverages blockchain technology to provide financial services without traditional intermediaries such as banks. By enabling peer-to-peer transactions through decentralized applications (dApps), DeFi promotes financial inclusion, lower transaction costs, and improved transparency. However, with these advantages come significant cybersecurity risks due to the underlying technology and its decentralized nature.

Key Security Risks in DeFi

1. **Smart Contract Vulnerabilities** Smart contracts are self-executing agreements encoded on blockchain platforms. Since they automate complex financial transactions, any flaw in the code can result in catastrophic financial losses. Many DeFi protocols use open-source code, making them susceptible to exploitation by malicious actors. For example, bugs or logic errors in smart contracts have led to numerous high-profile hacks, including the infamous DAO hack in 2016, where attackers exploited a vulnerability to siphon approximately $60 million in Ether[213].

2. **Oracle Manipulation** DeFi platforms rely on oracles to fetch real-world data, such as asset prices or exchange rates. If an oracle is compromised or manipulated, attackers can force the system to execute unfavorable trades. A well-known instance occurred in April 2020, when attackers

---

[213] https://www.gemini.com/cryptopedia/the-dao-hack-makerdao

exploited a vulnerability in the bZx protocol's oracle system, causing a loss of nearly $8 million[214].

3. **Flash Loan Attacks** Flash loans allow users to borrow funds without collateral, provided they repay the loan within a single blockchain transaction. While this innovation enhances liquidity, it has become a tool for market manipulation. In February 2020, the DeFi protocol bZx suffered another flash loan attack when hackers borrowed and manipulated asset prices through oracle tampering, resulting in a $350,000 loss[215].

4. **Rug Pull Scams** Rug pulls occur when DeFi developers create seemingly legitimate projects, attract significant investments, and then suddenly withdraw all funds, leaving investors with worthless tokens. One notable case was the 2020 Harvest Finance incident, where over $24 million was drained from the platform due to what appeared to be an insider exploit. Another example is the Squid Game Token scam, where developers vanished after collecting millions in investments[216].

Notable Case Studies

1. **The $600 Million Poly Network Hack (August 2021)** In one of the largest DeFi hacks in history, attackers exploited security flaws in the Poly Network platform, stealing over $600 million worth of cryptocurrency. Surprisingly, the attackers returned the funds, claiming their intent was to

---

[214] https://www.immunebytes.com/blog/bzx-protocol-exploit-sep-14-2020-detailed-analysis/

[215] https://www.mdpi.com/2076-3417/14/14/6361

[216]
https://sustainabledevelopment.un.org/content/documents/1126SD21%20Agenda21_new.pdf

expose critical vulnerabilities. This incident underscored the need for rigorous security measures in DeFi systems[217].

2. **Compound Protocol Bug (September 2021)** A bug in the Compound Finance protocol's upgrade led to unintended distribution of $90 million in COMP tokens to users. The flaw originated from an incorrect implementation of the smart contract upgrade. Although the team managed to recover some funds, the incident highlighted the risks of protocol upgrades without comprehensive testing[218].

Defensive Strategies

1. **Smart Contract Audits** Conducting regular security audits by reputable firms like CertiK, OpenZeppelin, and ConsenSys Diligence helps identify and fix vulnerabilities in smart contracts before deployment. Comprehensive audits involve code reviews, penetration testing, and formal verification.

2. **Bug Bounty Programs** Offering financial rewards to ethical hackers incentivizes them to discover and report vulnerabilities. Programs like Immunefi have become popular in the DeFi ecosystem, paying out millions in rewards to white-hat hackers.

3. **Insurance Protocols** DeFi insurance platforms like Nexus Mutual and Cover Protocol provide coverage against losses from hacks and system failures. While still a developing

---

[217]https://www.reuters.com/technology/how-hackers-stole-613-million-crypto-tokens-poly-network-2021-08-12/
[218]https://www.garp.org/hubfs/Whitepapers/a2r5d000003IhylAAC_RiskIntell.WP.DeFiRisks.7.29-1.pdf

field, DeFi insurance has gained traction, helping to mitigate risks for investors.

4. **Decentralized Governance** Implementing decentralized governance ensures community participation in decision-making processes. Protocols like MakerDAO, Compound, and Uniswap rely on governance tokens that allow stakeholders to vote on important proposals, reducing insider risks and enhancing transparency.

5. **Security Standards and Frameworks** Industry bodies like the Blockchain Security Alliance are working toward establishing security standards and best practices. Adhering to these guidelines can help DeFi platforms build more secure systems.

Industry Response

Major DeFi platforms such as Aave, Compound, and MakerDAO have adopted comprehensive security frameworks involving continuous auditing, community-driven governance, and multi-signature wallet implementations. Additionally, collaborations with cybersecurity firms and the introduction of decentralized insurance protocols have strengthened the ecosystem's resilience.

Global regulatory bodies are also paying closer attention to DeFi, exploring policies to manage associated risks while fostering innovation. In October 2022, the Financial Action Task Force (FATF) updated its guidance on virtual asset service providers (VASPs), emphasizing the need for anti-money laundering (AML) and counter-terrorism financing (CTF) measures in the DeFi space[219].

---

[219] https://www.fatf-gafi.org/content/dam/fatf-gafi/guidance/RBA-VA-VASPs.pdf

Conclusion
While DeFi offers unparalleled financial opportunities, its decentralized architecture presents unique security challenges. The sector's rapid growth necessitates constant vigilance, improved technical standards, and collaborative efforts among developers, security experts, and regulators. By implementing robust defensive strategies and learning from past incidents, DeFi platforms can build a more secure and sustainable financial ecosystem for the future.

📖 Conclusion

The FinTech sector's rapid digital transformation brings both unprecedented opportunities and evolving cybersecurity challenges. Quantum computing threatens traditional encryption, AI-driven attacks increase the sophistication of cyber threats, and DeFi introduces unique vulnerabilities. Staying ahead of these trends requires continuous innovation, collaboration, and a proactive cybersecurity culture. By adopting advanced defense mechanisms, conducting frequent security assessments, and fostering industry-wide partnerships, FinTech companies can mitigate emerging risks and ensure a secure digital financial ecosystem.

💡 What We Learnt:

- **Quantum Computing Threat:** Quantum computing could break current cryptographic systems like RSA and ECC, risking sensitive financial data.

- **Post-Quantum Cryptography (PQC):** New encryption algorithms like lattice-based cryptography are being developed to resist quantum attacks.

- **Industry Preparation:** Financial institutions like JPMorgan Chase and tech giants like IBM are investing in quantum-safe protocols.

- **AI-Driven Cyberattacks:** Cybercriminals are using AI for deepfakes, automated phishing, and adversarial machine learning attacks.

- **Defensive AI Solutions:** Organizations are deploying AI-powered threat detection, behavioral analytics, and threat intelligence platforms to combat AI-driven threats.

- **Case Study Insight:** The 2020 Twitter Bitcoin scam highlighted the role of AI bots in amplifying cyberattacks.

- **DeFi Security Risks:** Decentralized finance platforms face smart contract bugs, oracle manipulation, flash loan exploits, and rug pull scams.

- **Smart Contract Audits:** Regular audits by firms like CertiK can reduce DeFi vulnerabilities.

- **Bug Bounty Programs:** Platforms like Immunefi incentivize ethical hackers to report bugs.

- **Insurance Protocols:** DeFi insurance platforms offer protection against financial losses from hacks.

- **Global Regulation:** Regulatory bodies like FATF are introducing policies to manage risks in the DeFi ecosystem.

- **Overall Cybersecurity Imperative:** Staying ahead of cybersecurity threats requires continuous innovation, industry collaboration, frequent security audits, and the adoption of emerging security technologies.

CHAPTER 15
BEST PRACTICES FOR FINTECH SECURITY

The FinTech industry has revolutionized the way individuals and businesses manage their financial transactions, investments, and overall economic well-being. With the rapid advancement of digital technologies, FinTech companies have introduced innovative solutions that make banking, payments, lending, and wealth management more accessible and efficient than ever before. This unprecedented growth has reshaped global financial landscapes, enabling seamless, real-time financial services through web platforms and mobile applications.

However, with great innovation comes an equally significant need for robust security measures. As FinTech products handle sensitive financial data and facilitate critical business operations, they have become prime targets for cybercriminals. Threat actors continuously develop sophisticated tactics to exploit vulnerabilities, compromise confidential information, and disrupt financial ecosystems. Data breaches, ransomware attacks, and identity theft pose severe risks, potentially leading to financial losses, reputational damage, and regulatory penalties.

Given the high stakes, securing FinTech platforms is not just a technical necessity but a business imperative. Building trust with customers depends on creating secure digital environments where their data and transactions remain protected. Regulatory compliance adds another layer of responsibility, requiring companies to adhere to stringent security standards and privacy laws to operate legally in various regions.

This chapter delves into the essential security practices every FinTech company must implement to safeguard their systems against evolving cyber threats. We will cover strategies for securing sensitive data, strengthening authentication mechanisms, and ensuring business continuity in the face of potential attacks. Additionally, we will outline actionable takeaways to help developers, security teams, and business leaders integrate security measures into every stage of product development.

To make these concepts practical, we have included a detailed checklist designed to help FinTech teams evaluate and improve their security posture. From encrypting data and conducting regular vulnerability assessments to implementing multi-factor authentication and ensuring regulatory compliance, this checklist serves as a comprehensive guide for building resilient, secure FinTech products.

By the end of this chapter, readers will have a deeper understanding of cybersecurity challenges specific to the FinTech industry and the knowledge required to create products that prioritize both functionality and security. With a proactive approach, FinTech companies can navigate the complexities of the digital financial world while maintaining trust and safeguarding their customers' most valuable assets.

📖 Comprehensive Security Practices

a. Continuous Monitoring

Continuous monitoring is the cornerstone of FinTech security[220], serving as a critical mechanism for safeguarding sensitive financial data, ensuring compliance, and maintaining system integrity. It involves the real-time tracking of network activity, application performance, and data flows to identify and respond swiftly to potential threats before they escalate. By continuously analyzing system behaviors and user activities, organizations can detect anomalies, unauthorized access attempts, and vulnerabilities that may compromise their infrastructure.

The significance of continuous monitoring extends beyond mere detection; it also facilitates proactive incident response, allowing security teams to mitigate risks quickly and efficiently. Automated alerts, data analytics, and AI-driven insights play a crucial role in enhancing visibility across complex IT environments, enabling organizations to stay ahead of emerging cyber threats.

Without continuous monitoring, FinTech companies face increased exposure to various cyber risks, including data breaches, fraud, and service disruptions. Such incidents can lead to severe financial losses, eroded customer trust, and long-term damage to a company's reputation. In a sector where customer confidence and regulatory compliance are paramount, adopting a robust continuous monitoring strategy is not just an option—it's an essential component of a resilient cybersecurity framework.

---

[220]https://www.uptech.team/blog/fintech-security

Key Strategies:

- **Implement Security Information and Event Management (SIEM):** Use SIEM tools to centralize data collection, analyze logs, and generate alerts, helping security teams quickly identify and respond to potential threats.

- **Adopt Threat Intelligence Feeds:** Integrate external threat intelligence services to stay ahead of emerging threats by accessing real-time threat data and indicators of compromise.

- **Automate Alerts and Responses:** Use automated tools to detect anomalies and trigger predefined responses, minimizing manual intervention and response times.

- **Conduct Regular Audits:** Perform regular system audits and vulnerability assessments to ensure compliance and identify security gaps before they can be exploited[221].

b. Data Encryption and Protection

Data protection is a cornerstone of the FinTech industry, where organizations handle vast amounts of sensitive personal and financial information daily. A single breach can expose confidential data, leading to severe financial losses, reputational damage, and legal consequences for both businesses and individuals. In this context, implementing robust security measures is not just a regulatory requirement but a critical business necessity.

Among the various security protocols, encryption stands out as one of the most effective defense mechanisms. It works by converting sensitive data into an unreadable format that can only be decoded with the correct decryption key. This ensures that even if data is

---

[221]https://www.miquido.com/blog/security-in-fintech/

intercepted during transmission or compromised through a breach, it remains unintelligible and unusable to unauthorized parties. Encryption safeguards information such as bank account details, credit card numbers, and personal identification data, forming a protective barrier against cybercriminals.

By deploying advanced encryption algorithms, FinTech companies can enhance customer trust, comply with data protection laws such as GDPR and PCI DSS, and strengthen their overall cybersecurity framework. In a digital economy where financial transactions occur at lightning speed, encryption serves as a critical shield, ensuring that sensitive data stays secure, confidential, and protected from potential threats.

Key Strategies:

- **Use End-to-End Encryption:** Ensure that all data in transit and at rest is encrypted using industry-standard protocols such as AES-256, safeguarding sensitive customer information against interception.

- **Tokenization:** Replace sensitive data with tokens that have no exploitable value outside secure systems, reducing the risk of data exposure.

- **Data Masking:** Mask sensitive data during processing and testing to reduce exposure while allowing necessary functions such as analytics.

- **Secure APIs:** Ensure API communications are encrypted and authenticated, preventing unauthorized access and data breaches.

c. Identity and Access Management (IAM)

Controlling who can access systems and data is a critical component of minimizing security risks in the ever-evolving digital landscape. FinTech firms, which handle sensitive financial information and process high-value transactions, must implement robust Identity and Access Management (IAM) frameworks to safeguard their operations. An effective IAM framework ensures that only authorized individuals can access specific systems, applications, and data based on clearly defined roles and responsibilities. This approach reduces the risk of unauthorized access, data breaches, and insider threats. By enforcing the principle of least privilege, FinTech companies can ensure that employees, contractors, and third-party vendors have access only to the information necessary for their job functions. Additionally, regular audits, access reviews, and the integration of advanced technologies such as multi-factor authentication (MFA) and role-based access controls (RBAC) further strengthen the IAM framework, enhancing overall cybersecurity resilience.

Key Strategies:

- **Multi-Factor Authentication (MFA):** Require multiple verification factors such as passwords, biometrics, or device-based authentication for sensitive transactions and system access.

- **Role-Based Access Control (RBAC):** Assign permissions based on users' job functions, ensuring that employees only access the data and systems they need.

- **Least Privilege Principle:** Limit user permissions to only what is necessary for their tasks, reducing the attack surface and potential for insider threats.

- **Regular Access Reviews:** Conduct periodic reviews to ensure permissions remain appropriate as roles and responsibilities change within the organization.

d. Secure Product Development Culture

Building security into the product development lifecycle is essential for ensuring long-term resilience and protecting both the product and its users from potential threats. Security must be considered from the earliest stages of product design, starting with the conceptualization and planning phases. This includes conducting risk assessments, identifying potential vulnerabilities, and integrating best practices for secure coding.

As development progresses, implementing secure development frameworks, performing regular code reviews, and conducting thorough testing help to identify and mitigate security issues early. Security should not be an afterthought but an integral part of every development sprint and milestone.

Deployment is another critical stage where security measures such as secure configurations, encryption protocols, and access controls must be enforced. Ongoing monitoring, incident response planning, and regular updates ensure that the product remains resilient against evolving threats.

Ultimately, embedding security throughout the product development lifecycle fosters a proactive security culture, reduces risks, and builds customer trust by delivering a product that is secure, reliable, and prepared to adapt to future challenges[222].

---

[222]https://www.capitalnumbers.com/blog/fintech-cybersecurity-practices/

Key Strategies:

- **Secure Development Lifecycle (SDLC):** Integrate security checks at every stage of development, from planning and design to testing and release.

- **Code Reviews and Testing:** Conduct peer code reviews and use automated testing tools to identify vulnerabilities before they are released to production.

- **Secure Coding Standards:** Establish guidelines for writing secure code, focusing on preventing common vulnerabilities such as SQL injection and cross-site scripting.

- **Third-Party Component Management:** Vet and monitor third-party libraries and frameworks to ensure they are free from known vulnerabilities.

e. Incident Response and Recovery

Despite even the most robust preventive measures, security breaches can still occur due to evolving threats and unforeseen vulnerabilities. Having a well-structured incident response plan in place is essential to minimize the potential damage caused by such breaches. This plan ensures that an organization can respond swiftly and effectively, reducing downtime, mitigating financial losses, and preserving its reputation. A comprehensive incident response strategy includes clear protocols for identifying, containing, and resolving security incidents while safeguarding critical data and maintaining operational continuity. By acting quickly and transparently, businesses can not only recover more efficiently but also reinforce customer trust and confidence in their ability to handle unforeseen challenges.

Key Strategies:

- **Develop an Incident Response Plan:** Create a detailed plan outlining roles, communication protocols, and recovery steps to ensure a coordinated response.

- **Simulate Breach Scenarios:** Conduct regular breach simulation exercises to test the organization's preparedness and improve response times.

- **Data Backup and Recovery:** Ensure regular backups and establish disaster recovery sites to maintain business continuity during severe incidents.

- **Post-Incident Reviews:** Conduct post-incident reviews to identify root causes, learn from breaches, and prevent future occurrences[223].

📖 Key Takeaways and Checklist for Building Secure FinTech Products

a. Governance and Compliance

Staying compliant with regulatory standards is not optional—it is a legal obligation that organizations must take seriously to avoid legal repercussions, financial penalties, and reputational damage. In today's ever-changing regulatory environment, businesses must continuously monitor and adapt to evolving laws and industry-specific guidelines. This requires a proactive approach that includes establishing a dedicated compliance team responsible for tracking regulatory changes, assessing their impact on business operations, and ensuring adherence through regular audits and employee

---

[223]https://kindgeek.com/blog/post/developing-secure-fintech-application

training programs. By fostering a culture of compliance, organizations can mitigate risks, build trust with stakeholders, and maintain a competitive edge in the marketplace.

Checklist:
- Stay updated with relevant regulatory standards (e.g., PCI-DSS, GDPR, ISO/IEC 27001).

- Establish a dedicated compliance team.

b. Risk Management

A well-defined risk management framework helps identify, assess, and mitigate potential security risks.

Checklist:
- Conduct regular risk assessments.

- Develop a risk management framework tailored to your business model[224].

c. Employee Training and Awareness

Employees are often considered the weakest link in an organization's cybersecurity defenses due to the potential for human error, negligence, or lack of awareness about security protocols. Cybercriminals frequently exploit this vulnerability through tactics such as phishing scams, social engineering, and malware attacks. However, this risk can be significantly mitigated through comprehensive cybersecurity training programs designed to educate employees about potential threats and best practices.

By fostering a well-informed workforce, businesses can reduce the likelihood of breaches caused by avoidable mistakes like clicking

---

[224]https://moldstud.com/articles/p-security-in-the-digital-age-best-practices-for-fintech-developers

on suspicious links, sharing sensitive information, or using weak passwords. Regular workshops, simulated phishing tests, and clear communication about security policies help reinforce good cybersecurity habits. In addition, creating a culture of accountability and emphasizing the shared responsibility of protecting organizational data can strengthen overall security. An educated and vigilant team becomes an essential line of defense, transforming a potential vulnerability into a powerful asset in the fight against cyber threats.

Checklist:

- Conduct mandatory security awareness training.

- Implement phishing simulations and social engineering tests.

d. Customer Security Awareness

Customers play a crucial role in maintaining security, as their actions can directly impact the safety of their personal information and the integrity of the services they use. In today's digital world, where cyber threats such as identity theft, phishing scams, and account takeovers are on the rise, educating customers about safe online practices is more important than ever.

Businesses have a responsibility to raise awareness by providing clear, accessible guidance on cybersecurity best practices. This includes encouraging the use of strong, unique passwords, enabling multi-factor authentication, and recognizing signs of phishing attempts. Regularly sharing security tips through emails, websites, and mobile app notifications can keep customers informed and vigilant.

Interactive tutorials, webinars, and security awareness campaigns can further enhance customer knowledge, empowering them to protect themselves from cybercrime. Businesses can also establish support channels to assist customers in securing their accounts and responding swiftly to suspicious activities.

By fostering a partnership with customers based on trust and shared responsibility, organizations can significantly reduce the risk of fraud, data breaches, and unauthorized account access. An informed and security-conscious customer base becomes a valuable asset in the ongoing effort to maintain a safe and resilient digital environment.

Checklist:
- Educate customers on cybersecurity best practices.

- Provide secure account recovery mechanisms.

e. Security Testing

Frequent security testing helps identify vulnerabilities before attackers can exploit them.

Checklist:
- Conduct penetration testing and bug bounty programs.

- Perform security audits by certified third-party vendors.

f. Supply Chain Security

Vulnerabilities in third-party vendors can compromise the entire FinTech ecosystem, posing significant risks to both financial institutions and their customers. In an interconnected digital environment, FinTech companies often rely on external service providers for critical functions such as payment processing, cloud storage, data analytics, and software development. While these partnerships enhance operational efficiency and enable innovation, they also expand the organization's attack surface.

Cybercriminals frequently target third-party vendors because they may have weaker security controls compared to the financial institutions they serve. A single vulnerability in a vendor's system can create a backdoor, allowing attackers to infiltrate sensitive financial networks. This can lead to data breaches, financial fraud, and disruptions in essential services, undermining customer trust and damaging brand reputation.

To mitigate these risks, FinTech companies must adopt a robust vendor management strategy. This includes conducting thorough due diligence before engaging with third-party providers, evaluating their security protocols, and requiring compliance with industry standards such as SOC 2, ISO 27001, or PCI-DSS. Regular security audits, risk assessments, and penetration testing should be mandatory to identify and address potential vulnerabilities.

Additionally, establishing clear contractual agreements that outline cybersecurity responsibilities, incident response procedures, and data protection obligations is critical. Continuous monitoring of vendors' security postures can provide real-time visibility into potential threats, enabling faster response times in the event of a breach.

By prioritizing third-party risk management, FinTech companies can strengthen their defenses, ensuring a more secure and resilient financial ecosystem that protects sensitive customer information and upholds regulatory compliance.

Checklist:
- Vet vendors and partners for security compliance.
- Include security clauses in service-level agreements (SLAs).

📖 Final Words

Navigating the rapidly evolving cybersecurity landscape in FinTech requires a proactive, layered, and comprehensive approach. With cyber threats becoming more sophisticated, securing financial technology platforms is not merely a technical challenge—it's a business imperative.

By implementing the best practices outlined in this chapter, FinTech companies can build robust, secure, and resilient systems. Success lies in combining cutting-edge technologies, vigilant monitoring, and a strong security culture that permeates every level of the organization.

FinTech security is a continuous journey. Stay vigilant, stay prepared, and keep adapting to the evolving threat landscape to maintain trust and safeguard critical financial assets[225].

💡 What We Learnt:

- **FinTech Security Importance:** FinTech companies must prioritize security to protect sensitive financial data, maintain customer trust, and ensure regulatory compliance.

- **Cyber Threat Landscape:** The FinTech sector faces evolving threats such as data breaches, ransomware, and identity theft, requiring advanced security measures.

---

[225]https://www.cynance.co/cybersecurity-in-fintech-7-ways-to-stay-safe/

- **Continuous Monitoring:** Real-time monitoring helps detect and respond to threats proactively, reducing downtime and limiting potential damages.

- **Data Encryption and Protection:** Encrypting data ensures its confidentiality, making it unreadable to unauthorized users even if intercepted or compromised.

- **Identity and Access Management (IAM):** Controlling user access through multi-factor authentication and role-based permissions minimizes risks of unauthorized data exposure.

- **Secure Product Development Culture:** Security must be integrated throughout the product development lifecycle, from design to deployment, to build resilient systems.

- **Incident Response and Recovery:** Having a well-structured incident response plan ensures rapid response to breaches, minimizing damage and restoring operations quickly.

- **Governance and Compliance:** Staying updated with regulatory standards and conducting regular audits ensure legal compliance and reduce legal risks.

- **Risk Management:** A proactive risk management framework helps identify, assess, and mitigate potential cybersecurity threats.

- **Employee Training and Awareness:** Regular employee training reduces human error and strengthens the organization's first line of defense against cyberattacks.

- **Customer Security Awareness:** Educating customers on secure practices enhances their ability to protect their accounts and personal data.

- **Security Testing:** Regular penetration testing, audits, and vulnerability assessments help identify and address security flaws before attackers exploit them.

- **Supply Chain Security:** Evaluating third-party vendors and ensuring their compliance with security standards minimizes risks from external service providers.

- **Security as a Continuous Process:** FinTech security is an ongoing process that requires constant updates, monitoring, and adaptation to new threats.

APPENDIX A
GLOSSARY OF TERMS AND ACRONYMS

API (Application Programming Interface):

An API is a set of protocols and tools that enable different software applications to communicate with each other. It defines methods of interaction between various software intermediaries, allowing them to request and exchange data efficiently. APIs are widely used in web development, mobile apps, and cloud services to integrate functionalities like payment gateways, social media sharing, and data retrieval.

Blockchain:

Blockchain is a decentralized and distributed digital ledger that records transactions across multiple computers to ensure security, transparency, and immutability. Each transaction is stored in a block, linked to the previous one, forming a continuous chain. This technology powers cryptocurrencies like Bitcoin and has applications in supply chain management, healthcare, and digital identity verification.

DeFi (Decentralized Finance):

DeFi refers to a system of financial services built on blockchain technology, allowing users to perform transactions such as lending, borrowing, trading, and investing without traditional financial intermediaries like banks. DeFi platforms use smart contracts to automate and secure financial operations, offering global access to financial services with transparency and lower fees.

GDPR (General Data Protection Regulation):

The GDPR is a comprehensive legal framework established by the European Union to safeguard personal data and protect individual privacy. It sets rules for how organizations collect, store, and process personal information. Companies must ensure data security, provide transparency, and obtain user consent for data processing. Non-compliance can result in significant fines and penalties.

PCI DSS (Payment Card Industry Data Security Standard):

PCI DSS is a global security standard for organizations that handle credit card transactions. It establishes security protocols to protect cardholder data from breaches and fraud. Requirements include encrypted data storage, secure payment processing environments, and regular security assessments. Compliance is essential for businesses to maintain customer trust and avoid financial losses.

PSD2 (Payment Services Directive 2):

PSD2 is a European regulatory framework designed to enhance electronic payment services' security, competition, and innovation. It requires banks to provide third-party payment service providers access to customer account information (with consent) through APIs. This opens the financial market to fintech innovations, promoting transparency and better services for consumers.

Two-Factor Authentication (2FA):

2FA is a security measure that requires users to provide two forms of identification to access an account or perform sensitive operations. Typically, this involves a combination of something the user knows (password), something they have (a security token or phone), or something they are (biometric data like a fingerprint). 2FA strengthens account security by adding an extra layer of protection against unauthorized access.

ABOUT THE AUTHOR
HASHNEE SUBBUSUNDARAM

Hashnee is a highly accomplished and renowned Cybersecurity Program Lead with over 15 years of experience driving transformative initiatives across Fortune 500 companies and high-growth enterprises. As one of the few women in cybersecurity, she is a trailblazer in an industry traditionally dominated by men, advocating for women in STEM and setting a high standard for excellence.

Throughout her career, Hashnee has held pivotal roles at leading organizations such as Visa, Qualcomm, Cisco, LendingClub, and San Mateo County. At Visa, she led the creation of groundbreaking tools that mitigated high-risk events like fraud and cyberattacks across 200 countries, setting a new global security benchmark. Her leadership has consistently delivered significant cost savings, enhanced security postures, and improved operational efficiency, making her a trusted authority in cybersecurity.

Holding advanced degrees from prestigious institutions like Georgia State University (MS), Bharathiar University (MBA), and Anna University (BE), Hashnee seamlessly combines business insight with technical expertise. She also serves on the Advisory Board for the Women in Leadership Program at the University of Colorado, Colorado Springs (UCCS). Passionate about diversity, Hashnee mentors young women and underrepresented groups, inspiring the next generation of cybersecurity leaders.

ABOUT THE AUTHOR
BALASUNDARAM SUBBUSUNDARAM
(BALA)

*Bala is a visionary FinTech leader with nearly two decades of experience delivering AI-powered financial solutions that set new industry standards. As Head of Product for Payments & Financial Services at Walmart Inc., a Fortune 1 company, he leads the development of embedded financial products and digital payment ecosystems that maximize customer lifetime value and operational efficiency. Bala is also the celebrated author of the acclaimed book **Fly High with AI**, a visionary masterpiece offering a definitive guide to launching a product management career in the age of artificial intelligence, praised for its profound insights and widespread acclaim.*

Bala's career spans impactful roles at LendingClub, ESM Solutions, TurningTech, and Taulia (an SAP company). At LendingClub, he spearheaded the company's transformation into a neo-bank, expanding investment portfolios and enhancing borrower and investor engagement. His leadership on key projects has consistently driven business growth, operational stability, and market expansion, positioning him at the forefront of financial innovation.

In addition to his corporate success, Bala advises on Engineering Leadership at California State University, Chico, and supports the Transformational Leadership Program at Seton Hall University, helping shape future leaders. With advanced degrees from Syracuse and Anna University, he seamlessly integrates technical expertise with strategic foresight, fostering collaboration and innovation to position organizations for sustainable growth in a dynamic, competitive landscape.

www.ingramcontent.com/pod-product-compliance
Lightning Source LLC
Chambersburg PA
CBHW071324210326
41597CB00015B/1342